MW00643363

THE COMPLETE GUIDE

to Raising Pigs

Everything You Need
to Know Explained Simply

BY CARLOTTA COOPER

The Complete Guide to Raising Pigs: Everything You Need to Know Explained Simply

Copyright © 2011 by Atlantic Publishing Group, Inc.
1405 SW 6th Ave. • Ocala, Florida 34471 • 800-814-1132 • 352-622-1875–Fax
Web site: www.atlantic-pub.com • E-mail: sales@atlantic-pub.com
SAN Number: 268-1250

No part of this publication may be reproduced, stored in a retrieval system, or trans-mitted in any form or by any means, electronic, mechanical, photocopying, recording, scanning, or otherwise, except as permitted under Section 107 or 108 of the 1976 United States Copyright Act, without the prior written permission of the Publisher. Requests to the Publisher for permission should be sent to Atlantic Publishing Group, Inc., 1405 SW 6th Ave., Ocala, Florida 34471.

Library of Congress Cataloging-in-Publication Data

Cooper, Carlotta, 1962-
 The complete guide to raising pigs : everything you need to know explained simply / by Carlotta Cooper.
  p. cm.
Includes bibliographical references and index.
ISBN-13: 978-1-60138-379-2 (alk. paper)
ISBN-10: 1-60138-379-7 (alk. paper)
1. Swine. I. Title.
SF395.C774 2011
636.4--dc22
                    2011005543

LIMIT OF LIABILITY/DISCLAIMER OF WARRANTY: The publisher and the author make no representations or warranties with respect to the accuracy or completeness of the contents of this work and specifically disclaim all warranties, including without limitation warranties of fitness for a particular purpose. No warranty may be created or extended by sales or promotional materials. The advice and strategies contained herein may not be suitable for every situation. This work is sold with the understanding that the publisher is not engaged in rendering legal, accounting, or other professional services. If professional assistance is required, the services of a competent professional should be sought. Neither the publisher nor the author shall be liable for damages arising herefrom. The fact that an organization or Web site is referred to in this work as a citation and/or a potential source of further information does not mean that the author or the publisher endorses the information the organization or Web site may provide or recommendations it may make. Further, readers should be aware that Internet Web sites listed in this work may have changed or disappeared between when this work was written and when it is read.

TRADEMARK DISCLAIMER: All trademarks, trade names, or logos mentioned or used are the property of their respective owners and are used only to directly describe the products being provided. Every effort has been made to properly capitalize, punctuate, identify and attribute trademarks and trade names to their respective owners, including the use of ® and ™ wherever possible and practical. Atlantic Publishing Group, Inc. is not a partner, affiliate, or licensee with the holders of said trademarks.

Printed on Recycled Paper

BOOK MANAGER: Amy Moczynski • AMoczynski@atlantic-pub.com
BOOK PRODUCTION DESIGN: T.L. Price • design@tlpricefreelance.com
PROOFREADER: Brett Daly • brett.daly1@gmail.com
COVER DESIGN: Meg Buchner • megadesn@mchsi.com
COLOR INSERT & BACK COVER DESIGNS: Jackie Miller • millerjackiej@gmail.com
Printed in the United States

We recently lost our beloved pet "Bear," who was not only our best and dearest friend but also the "Vice President of Sunshine" here at Atlantic Publishing. He did not receive a salary but worked tirelessly 24 hours a day to please his parents. Bear was a rescue dog that turned around and showered myself, my wife, Sherri, his grandparents Jean, Bob, and Nancy, and every person and animal he met (maybe not rabbits) with friendship and love. He made a lot of people smile every day.

We wanted you to know that a portion of the profits of this book will be donated to The Humane Society of the United States. *–Douglas & Sherri Brown*

The human-animal bond is as old as human history. We cherish our animal companions for their unconditional affection and acceptance. We feel a thrill when we glimpse wild creatures in their natural habitat or in our own backyard.

Unfortunately, the human-animal bond has at times been weakened. Humans have exploited some animal species to the point of extinction.

The Humane Society of the United States makes a difference in the lives of animals here at home and worldwide. The HSUS is dedicated to creating a world where our relationship with animals is guided by compassion. We seek a truly humane society in which animals are respected for their intrinsic value, and where the human-animal bond is strong.

Want to help animals? We have plenty of suggestions. Adopt a pet from a local shelter, join The Humane Society and be a part of our work to help companion animals and wildlife. You will be funding our educational, legislative, investigative and outreach projects in the U.S. and across the globe.

Or perhaps you'd like to make a memorial donation in honor of a pet, friend or relative? You can through our Kindred Spirits program. And if you'd like to contribute in a more structured way, our Planned Giving Office has suggestions about estate planning, annuities, and even gifts of stock that avoid capital gains taxes.

Maybe you have land that you would like to preserve as a lasting habitat for wildlife. Our Wildlife Land Trust can help you. Perhaps the land you want to share is a backyard—that's enough. Our Urban Wildlife Sanctuary Program will show you how to create a habitat for your wild neighbors.

So you see, it's easy to help animals. And The HSUS is here to help.

THE HUMANE SOCIETY
OF THE UNITED STATES.

2100 L Street NW • Washington, DC 20037 • 202-452-1100
www.hsus.org

Photo courtesy of USDA Natural Resources Conservation Service.

AUTHOR
Dedication

This book is dedicated to my grandmother, Addie McNabb, a farmer's wife, and to my parents, Carl and Betty Cooper. They were the first people who taught me how to care for animals. I miss them every day.

AUTHOR
Acknowledgements

I would like to acknowledge the wonderful information provided by the Extension Service of the USDA. I read countless pamphlets and printouts about all aspects of raising pigs put out by various state Extension Services. They provided answers to so many questions and always had the most up-to-date information about agricultural methods. I would encourage anyone thinking of raising pigs to get in touch with their local county Extension Service agent. They can provide you with plenty of current information for your area.

I offer my deepest thanks to my friends Donna Fefee and Desterie Grimes for being with me through each step of this book. Without their support and willingness to listen to me when I needed to talk, this book would not exist. I know they both learned more about pigs than they ever dreamed possible!

Table of Contents

Chapter 3: Caringfor Young Pigs... 63

Chapter 4: The Birdsand the Bees 87

Chapter 5: Housing: Straw, Sticks, or Bricks? 119

Chapter 8: Herd Management 189

Chapter 9: Going to the Show......199

Table of Contents

Chapter 10: This Little Piggie Went to Market 225

Chapter 11: The Smokehouse 259

Introduction

If you are reading this book, you are probably considering some form of pig ownership, whether you intend to breed them, raise them for the market, or just keep a pet pig or two around the house. If that is the case, allow me to offer my congratulations: Not only are you taking the first step in a long, rewarding journey that can bring you profit, fun, and many rewarding life experiences, but you also hold in your hands an invaluable tool that will guide you every step of the way. A book of this type has a rare opportunity to "delight and instruct," as the Roman poet Horace said that all literature should do. These pages will include many fun facts about pigs, anecdotes, and other material to entertain you as you read. We will review highlights of the history of the pig and its relationship with mankind, along with its development into a major part of the world's food production.

At the same time, the book's main objective is to provide the practical information you need to get your operation under way, right away. You will learn about the different breeds for pets and meat, how to select a breed for your own purposes, and how to care for these delightful animals. In addition, you will learn how to go about selecting a farm from which to buy your stock, along with what to look for in the young pigs, their parents, and the farm itself.

This book will describe different types of structures used to house pigs and help you select the housing approach suitable for your plans. In addition, you will learn how to construct an appropriate structure or adapt existing structures to keep your pigs safe, healthy, and comfortable.

You will also find a detailed description of feeding practices, and because there are numerous approaches and schools of thought on this subject, you will learn which method best suits the style of pig rearing you intend to do. A pet pig can be fed almost entirely on table scraps, although such food requires supplementation and preparation to be healthy. You will learn about commercial feeds, supplements, preparing slop, and even how to grow some of the pig's food yourself.

The ultimate goal of all pig farmers is to raise happy, healthy pigs. From start to finish, the words of wisdom contained within this book will help you establish your herd in as little time as possible. Although there are no real shortcuts, you can avoid the common pitfalls and grow your own successful pig business.

CHAPTER 1 :

A Pig Primer

Prehistoric Paintings to Pork Barrel Politics

In the 19th century, a young girl led her father, the amateur archaeologist Marcelino Sanz de Sautuola, to a cave near the Spanish town of Santillana del Mar. There, on the cavern walls and ceilings, was an astonishing series of paintings depicting the animals known to Cro-Magnon humanity, including wild boars. Uranium-thorium dating of the pigments in these cave drawings lead some scientists to estimate that the images were 25,000 to 30,000 years old. Indeed, few animals have had longer and more extensive relationship with humanity than pigs, which were among the first animals to be domesticated. The earliest domesticated pigs, which descended from Eurasian wild boars, probably occurred in Central Asia about 10,000 years ago. By 5000 B.C., the practice of keeping pigs was widespread, and Emperor Fo-Hi wrote the first book on raising pigs in 3468 B.C., the same year he is thought to have also penned the traditional book of prophecy, the *I-Ching*. Zhou period tombs from ancient China

(1121–221 B.C.) often included pigs carved from precious stones, thought to ensure the deceased prosperity in the afterlife.

In Europe, it is estimated that pigs were first domesticated around 5000 B.C., and the animal played an important role in much of the history and mythology of the continent. In the *Aeneid*, Virgil claimed that in the 6th century B.C. Aeneas saw a vision of a "sacred white sow," the female pig indicating the place on the Tiber river that should be the site of Rome. Pork was the favorite meat of the Romans, who considered cows to be beasts of burden and did not often eat veal or beef. In Petronius' *Trimalchio's Banquet*, roast boar with dates was the centerpiece of the feast, and such dishes were a central part of the wealthy Roman diet. If, as Napoleon said, "an army marches on its stomach," the Roman Empire would be built upon a foundation of bacon, as its legions were rationed with pork, grain, and wine; additionally, the boar was sacred to Mars, the Roman god of war.

Among the Celts, the boar was seen as a symbol of fertility, power, and prosperity, and the animal was sacred to the goddess of the hunt, Arduinna. The bones and joints of pigs have been discovered buried in Celtic tombs, suggesting the importance of these animals in their culture. The Druids referred to themselves as boars and emulated the animals' solitary forest existence. Throughout the Middle Ages, the importance of pigs to the developing agrarian economy grew steadily, and the boar appears in many examples of heraldry.

In 1493, Christopher Columbus arrived on the island that would come to be known as Puerto Rico with eight pigs that Queen Isabella had ordered him to bring. Almost 50 years later, Hernando de Soto landed in Florida with a small herd of pigs that would grow to many hundreds in just a few years. De Soto and other Spanish explorers offered some of these animals to the indigenous peoples as a token of goodwill, and numbers of them escaped into the wild, becoming the ancestors of the razorbacks and feral pigs of the southern United States.

Sir Walter Raleigh brought a number of sows to the Jamestown colony in 1607, and though the settlers endured a great deal of hardship, they were to be the first permanent English establishment in the New World. Over the course of the next century, pigs were to become a staple throughout the colonies because of their natural adaptability and utility.

As pioneers moved west, they brought pigs with them in their wagon trains, and soon large herds were to be found across the settled regions. In the War of 1812, salt pork was shipped to American troops in barrels stamped "U.S." Although the "U.S." stood for the meat packer "Uncle" Samuel Wilson, soldiers coined the term "Uncle Sam," one of America's most widely recognized patriotic symbols. Throughout the history of the United States, the importance of hogs to the nation's economy has continued to grow, along with the relationship between pigs and politics. In the 19th century, the term "pork barrel politics" was coined to describe actions undertaken by politicians to solely benefit their constituencies. The "pork barrel" was a container that held lard or salt pork, and keeping it well supplied was a matter of constant concern for the average American.

In 1961, pigs broke into the financial markets when the Chicago Mercantile Exchange began trading pork belly futures contracts as an innovative risk management device for meat packers; because frozen pork bellies can be kept in cold storage for extended periods, they can be held in inventory and sold when market prices meet the sellers' needs. Pigs are no less important elsewhere in the world; today, every continent has a pig population, except Antarctica.

Are You Ready to Raise Pigs?

Pigs are wonderful, intelligent, sociable animals, but it is important that you seriously consider your ability to live up to the responsibilities that proper care for them entails. Although the decision to keep an animal should never be taken lightly, this is particularly true of pigs, which have very special requirements — though this is counterbalanced by the rewarding experience you can have with them. To begin with, if you are thinking of keeping a pig as a pet, you should be aware that the average lifespan of a domestic pig is ten to 15 years, and some breeds can live substantially longer.

The next important consideration is the size of animal you can handle. In the 1990s, it became very trendy to own a "yuppy puppy," or miniature pig, but many people did not realize that "miniature" is a relative term when applied to pigs. When the full-sized animal can be 900 pounds, or even bigger, miniature can mean 200 pounds. As a result, many pigs ended up in animal shelters waiting their turn to be euthanized. The problem persists today, though numerous pig rescue organizations have formed in response. By working to educate others about pigs and by working hard to ensure your pigs are properly cared for, you can make yourself a part of the solution. You will learn more about this later in the book in the sections on breed selection and buying your first pig. *See Chapter 3.*

In terms of intelligence and curiosity, a pet pig is much like a toddler. If there is some trouble it can get into, it will. If there is something it should not eat within its reach, it will eat it. You can train your pet pig in the same way that you train a puppy, but if your pig is bored or left alone too much, it will eat makeup that is lying around, medicine, or just about anything else that smells remotely appetizing to it. Just as you would need to childproof your home when a toddler starts to walk, you need to pig-proof if you plan on having an indoor pig. If there is a way your cupboards can be opened or if there is a weakness in the fence, your pig will find it.

Depending on the size and type of pig, you may need to provide only a pet bed or a nest of blankets and a pet door in your house, or you may need to establish an entire indoor/outdoor area for its use. *See Chapter 5 for more information on this.* To a great extent, the amount of space at your disposal for your pig's use determines the type of pig ownership that will be possible for you.

Another point that you will find stressed throughout this book is that pigs are social animals with emotional needs. It would be cruel to raise a pig in isolation. Make certain that your pig has plenty of companionship, whether it is other pigs, different animals, or yourself. Just as a pig is comparable to a toddler in other respects, you will probably need to devote a similar amount of time to caring for your pig as you would a child. Very few pet boarding facilities are set up to handle pigs; if you intend to travel, you should check to see if there is a boarding facility in your area that has the facilities and abilities to care for swine. If you plan to maintain a herd, you will probably have to give up the idea of family vacations for a while. Raising livestock requires constant attendance to administer their needs, monitor their health, and remove manure. Even pastured pigs require too much attention to be left alone for long.

Do not be discouraged: raising pigs is fun, rewarding, and profitable. Like any undertaking, though, it is important to start with your eyes open so you know what you are getting into. To this end, it will be better if you read this book through before you actually purchase an animal.

Characteristics and Behavior

*"No man should be allowed to be president
who does not understand hogs."*

Harry S. Truman

Domestic pigs have changed a great deal over the last century. The dwindling use of lard in cooking has led pork producers to breed a leaner animal to satisfy consumer demand. The very round, fat pigs of the past, called "chuffy" pigs or "cob rollers," are no longer as prevalent. Wild pigs, which are termed "leggy" or "rangy," do not quite meet the needs of the modern pig breeder either because they lack the meat quality consumers want. Instead, the contemporary swine falls somewhere in between the two, favoring a larger loin eye area.

A pig's anatomy is remarkably similar to that of a human being. It has a respiratory, cardiovascular, and digestive system so much like a human that it is common for them to be used as test animals in laboratories. This is very important because it helps pig owners understand these animals and how to care for them. Essentially, pigs need the same things humans do. If your diet, climate, and shelter seem comfortable to you, it is safe to assume it is suitable for your pig as well.

Like human beings, hogs are omnivores, which is probably a big part of their success, both as a species and as a complement to humanity. Omnivores are extremely adaptable, which allows them to survive more readily in new environments and ecosystems. Like many omnivores, pigs are extremely intelligent, using their exceptional brainpower to hunt out new food sources and opportunities. According to the National Pork Producers Council, pigs are the fourth-smartest animal after humans, apes, and dolphins.

Often, they are friendly, sociable creatures, but at the same time, they tend to be individualistic — each pig has a unique characteristic. Some pigs, especially **boars** (adult males), may exhibit very aggressive or anti-social tendencies, though most are gregarious, curious, and playful. Most males are castrated when they are a few days old because they are destined to be feeder pigs instead of being used for breeding. If you are considering raising pigs, you should raise them in groups or at least with some form of companionship. Most pigs bond readily with other animals or human beings, and you will want to ensure your pig is not lonely. Often, a badly behaved pig is a bored or lonely pig so you should attend to the animal's emotional needs — not only for the pig's sake but also because a contented pig is easier to manage. You should provide a pig with toys because pigs are often as playful as puppies. Suggested toys include empty plastic trash cans they can push around, a bowling ball, or knotted pieces of rope or strips of cloth attached to the walls of their pen, which they like to pull.

Pigs have very poor eyesight, as their small, bleary eyes seem to suggest. For this reason, they tend to be wary of new places and are easily startled. Although their wild counterparts can be nocturnal, modern domestic pigs lack the tapetum lucidum, or inner-eye reflective tissue, that allows improved night vision, so pigs are not fans of dark places. The placement of their eyes favors lateral vision, but they lack the fine musculature needed for sharp focusing. This means if you are trying to move pigs, you will need to consider the effect that darkness and their limited forward vision has on them. A pig may put up a terrible fight about moving forward into a dark building, for example, because its forward vision makes the area ahead seem frightening. But, if you use a carrier or scoot the pig forward by means of a chute with pig boards placed close behind it and in front of it, the pig will move forward without any problem.

Pigs tend to interact with each other through their excellent sense of hearing and ability to vocalize, making pigs great communicators. It has been observed that pigs can make more than 20 distinct sounds that communicate meaning to other pigs. For instance, a single grunt, short or long, seems to indicate that a pig is happy, while a pig that grunts many times in succession is hungry. When a pig grunts several short times together, it normally means the pig is angry, and when a pig squeals, it is an indication the pig is in pain or fear. **Sows** — adult female pigs — give instructions to their piglets at feeding time and are known to make soothing noises as their babies nurse.

For all the power of pigs' ears, their most important tool is their super sensitive snout. A pig's snout is its fingers, eyes, and nose, all in one. The surface of a pig's snout is covered in thousands of tiny hairs that aid in capturing scents. Not only can they smell out all kinds of different foods with that powerful snout, but it also contains a structure of cartilage that they use as a sort of shovel to root around, turning over clumps of earth to expose tender plants or grubs. They achieve a surprising degree of dexterity with their snouts, rivaling even the elephant's remarkable "fingers" in the tip of its trunk. Although a well-fed pig does not need to root to find food, most will do so to amuse themselves and better understand their environment so it is best to provide them with fresh straw or hay. Sometimes rooting can be an undesirable behavior, and some farmers put a humane ring through a pig's nose in order to prevent it. If you are keeping pet pigs, be aware that most breeds will tear up lawns or flowerbeds. Rooting is an excellent way to turn the soil, but it can destroy anything you planted.

Pigs have rather short legs when compared with the length of their bodies, although this does not prevent young or fit pigs from achieving surprising speeds — in some places pigs are even raced. Their feet have four toes, each with an individual hoof, though they walk only on their front two, which are larger and more solid, while the other two dewclaws at the back of the hoof rarely touch the ground, except when a pig moves at top speed.

One of the most unfair stereotypes about pigs is that they are dirty animals. It is certainly true that pigs like to wallow, or lie about in mud and water, because they do not have sweat glands. A cool wallow allows them to regulate their body temperatures when it is hot. In addition, light-colored pigs are susceptible to sunburn and biting insects, and a coat of mud will protect them from both. But, if your pigs are given sufficient shade and a clean living area free of fecal matter to minimize insects, your pigs can wallow just as well in fresh, clean water. Most pigs are excellent swimmers, although you should know that potbellied breeds sometimes injure their bellies by kicking their back feet. You can provide your pigs with a mud wallow, a child's wading pool, or even a small, clean concrete pool. Pigs love water, and you should always make certain they have an ample supply for swimming and wallowing during the summer months, as well as for drinking.

Although most farm animals will defecate or urinate wherever they happen to be when the urge takes them, pigs do not do this. In fact, they are quite fastidious about such things, designating a specific place in which to do their business that is well separated from their nesting and feeding areas. Pigs living communally will also cooperate in this, establishing one or two restroom areas respected by all. This makes housetraining a pet pig very easy. Researchers at the University of Illinois also found that pigs will not play with a toy that has been soiled with feces.

Pigs are often found in dirty, muddy pens, but you should not keep your pigs this way. One reason people may do this is that it helps create emotional distance from animals intended for butchering. As a responsible owner, it is important that you make certain your pigs spend most of their lives with clean bodies, bedding, and food, even if a mud wallow is in the mix. It can be difficult to include a clean mud wallow in a pigpen but not impossible. Along with kiddie pools for your pigs, you could include a shallow concrete wallow filled with water. Not only will keeping your pigs clean make life better for the pigs, but it will also lead to better meat and a better overall experience for you, although it does mean more work.

Terminology

As with other animals, pigs have their own unique terminology. Anyone interested in raising a pig should get to know a few terms. In fact, even if you are simply interested in better understanding cuts of pork, it is helpful to learn some pig terminology.

Barrow	a young male pig
Boar	an adult male pig
Butcher–hog	a hog weighing between 220 and 260 pounds, raised for slaughter
Creep	the area in a farrowing site that is off-limits to the sow, where the piglets can get feed
Crossbred	a pig that is a cross of different purebred pig breeds
Dam	mother
Farrow	for a pig to give birth
Farrow to finish	raising pigs from birth to being market or slaughter-ready
Feed efficiency	the amount of feed it takes for a pig to gain 1 pound
Feeder pig	a young pig, usually just weaned, that is produced by a breeder but raised by someone else; usually between 35 and 70 pounds
Finished hog	a fattened hog that is ready for market or slaughter
Gilt	a young female pig
Grower pig	a young pig being raised for market or for slaughter; usually a pig more than 50 pounds
Grower to finish	raising pigs from the weaner stage (about 8 weeks) to market or slaughter size
Hog	a pig weighing more than 120 pounds
Market hog	a pig that weighs 220 to 260 pounds and could be sold at market
Purebred	a pig that belongs to a recognized purebred of pig
Shoat	a pig from weaning age to 120 pounds
Sire	father
Sow	an adult female pig
Swine	generic term for pigs
Topline	the spine or back or the pig, especially in silhouette
Underline	the line formed along the stomach of the pig

Anatomy and Physiology

Pigs have a cardiovascular system with a heart and lungs located in approximately the same places as a human's heart and lungs. They have a gastrointestinal tract similar to humans as well. As farm animals go, they are not the most efficient grazers, having only one stomach, unlike the cow that has four stomachs.

Pigs have a long snout that they can use like a digging tool when they "root." Rooting allows them to dig and turn over soil to find food or anything edible beneath the topsoil. As you might imagine, the pig also has a remarkable sense of smell that they use to sniff out anything to eat. Between their hooves and their very active snouts, pigs can keep the ground dug up very effectively.

The Mulefoot pig, a very rare old breed, has a single hoof on each foot like a mule. They are the only pig with this trait, although this trait is sometimes passed along when these pigs are crossbred.

Many people are unaware that pigs have hair. Their hair can range from a light, soft fluff to a harsh bristle, depending on the breed and sex of the pig. The color also varies a great deal. Pigs can be black, white, **piebald** — black with white bands or spots and points (feet and ears) — red, sandy, spotted, striped, gray, and mixes of these colors. Many people believe that pigs are pink, but "pink" pigs are actually white breeds. Their pink skin shows through, however.

Some breeds of pigs also have wattles. **Wattles** are long pieces of flesh that hang from their cheeks. They are not jowls but hang behind the jowls, under the ears. There is no real purpose for wattles. They are simply a feature on some breeds.

Pigs can have either erect or flop ears. Flop ears, also called lop ears, are drooping ears. Some ears are so long that they can cover the pig's eyes. In many cases, this is a good way to identify a breed if all other characteristics are the same.

So-called "hogzillas" that have been in the news in recent years in the United States have been found in Georgia and other Southern states. They are likely domestic pigs that have gone feral, according to DNA testing. Pigs can naturally revert to a feral state very rapidly as a result of a lack of interaction with humans. When this happens, their physical appearance can change, which includes a change in their skull. The dish-shaped face of the farmer's pig can change in just a few months to the straighter skull of the wild pig. This change allows the feral pig to use its snout more effectively for rooting and digging. The feral pig's teeth will continue to grow, particularly its wolf teeth that will become tusk-like. The feral pig will also continue to put on weight until it weighs far more than the farmer's pig.

On the farm, feeder pigs are normally slaughtered when they reach 230 to 250 pounds. It takes about five or six months after weaning to reach this weight. Sows kept on the farm for breeding may reach 600 pounds, depending on the breed. Boars kept for breeding may reach 800 pounds, depending on the breed. Some of the feral hogs discovered have weighed 800 pounds or more.

Without a trip to the butcher, the lifespan of a pig is about ten to 15 years, though some miniature pigs may live longer.

Summary

In this chapter you have looked at the pig from ancient times to the present. You have considered the pig's contribution to various cultures and how the pig has been perceived around the world. From a companion to gods to its association today with pork barrel politics, the pig has always represented bounteous prosperity.

You have also considered pigs as pets and some of the characteristics of pigs, in terms of both their personalities and their physical traits. Pigs are sometimes kept in dirty conditions, but they have no desire to be dirty animals. They are intelligent, playful animals that enjoy social interaction.

In the next chapter, you will begin to consider raising pigs. You will start to look at information about how to keep your pigs if you are interested in raising pigs for food. You will consider confinement methods, such as raising your pigs indoors or in pens or keeping your pigs on pasture. You will also examine organic pig raising, choosing your pigs, and the different pig breeds.

There is much to learn about raising pigs in the chapters ahead.

CHAPTER 2 :

Let's Get a Pig!

Now that you have looked at the pig's role in the world and its characteristics, it is time to consider the practical points of raising pigs today. As with most things, careful planning will be the key to success. The first and most important step is to determine the parameters that define success for you. What are your goals? Do you just want a miniature pet to be your pal? Are you interested in breeding miniature pigs to show or sell? On the other hand, maybe you are interested in producing meat for your table or the market. You may find that your experiences ultimately lead you in new directions, such as selling breeding stock or developing champion lines.

Another important consideration is the values that you bring to the project. Are you interested in contributing to the development of organic or sustainable markets? Are you looking for the best bottom line, the most enjoyable experience, or just a new hobby? If you are thinking of selling specialty meats based on a heritage breed or if you plan on marketing organic meat, will the market in your area support these activities?

If you are going to be successful in your pig-raising venture, you must be careful not to get in over your head: Do not move too quickly, be methodical, and use all the information available when making your decisions. Start with just one or two **table pigs**, or pigs you will raise for your own table or freezer, using the first year to get to know the animals and your aptitude for husbandry. You may decide that raising a couple of table pigs each year is quite enough for you, or you may want to expand your operation into breeding, raising for the market, or showing.

Confinement

Confinement is the most common method of pig farming used in commercial operations. It involves a large initial investment in constructing a housing unit and presents its own set of problems and advantages. In general, commercial confinement farms house 50 pigs or more in a single unit, which can have serious health implications. Outbreaks of disease can spread quickly in confinement housing, and ventilation may be inadequate in some cases. In addition, pigs raised in this way tend to be unhappy and take out their frustrations on each other, which is the reason that commercially raised pigs generally have their tails docked to prevent pigs from biting each other's tails.

On the other hand, confinement is very effective from an economic standpoint. It permits absolute control of the pigs' environment and requires less daily labor to maintain. Pigs in confinement can be fed on more intensive schedules, allowing remarkable feed-to-gain efficiency, with pigs growing at a maximum rate per pound of feed fed. Most pork in the United States is produced by commercial operations that use confinement farming. It allows producers to produce a large supply of pork and keep prices to consumers low. Still, full-time confinement farming is not for the beginner, nor is it appropriate for someone starting out with a couple of table or market pigs. If you are just starting out, this is not the way to do it.

Modified Confinement

A modified confinement system applies some of the principles of confinement but is applicable for smaller operations, as well as large operations. Unlike pigs in total confinement farms, which almost never see the outdoors, those in modified systems split their time between indoors and outdoors. To start a modified confinement system, you would need to construct a permanent housing unit and pen. Pigs raised in this fashion tend to be healthier and happier than those raised in total confinement, and this is the method most commonly used when raising just one or two at a time.

Modified confinement allows close control of the pigs' feeding schedules and other matters, while allowing the animals more access to fresh air, sunlight, and exercise. In addition, this is a good method when space is limited or when you do not have sufficient pasturage. For each pig you raise, you should have 6 to 8 square feet inside and out.

A modified confinement system can allow a farmer to pick and choose elements of raising pigs that fit the situation. The housing unit can be as large or as small as desired. A small number of pigs can be raised together, or the space can be expanded to contain a larger number. Elements of organic farming can be incorporated. The system can work for many different breeds of pigs and will work whether you raise feeder pigs for market or keep pigs as breeding stock.

Organic Farming

The methods used in organic swine farming differ considerably from those common to commercial farms. Organic farming is based on the belief that using natural processes will develop nutritionally superior food, as well as production methods that have a less negative environmental impact. Organic growers encourage microbial and bacterial diversity in their soil for the natural nutrients they impart to produce. Organic livestock production involves specialized feed and living conditions that are both more rewarding and challenging than standard methods. Organic methods are only possible if you have ample space for crops, if you intend to grow your own organic pig feed, and if you handle waste treatment. There is a growing market for organic food, but there are strict USDA regulations governing its production, and the products must be certified in order to be sold as organic. It is possible to use organic practices without gaining certification, however. If you use organic practices but your farm is not certified organic, you may wish to indicate this on the labeling of your pork products. Many consumers will still be interested in purchasing pork produced "naturally," that is "pasture raised," or raised "holistically," if you indicate that this is how the food has been produced on the packaging.

Livestock and meats that are sold as organic must fit an established set of criteria. To begin with, an animal's parentage must be known, and you must be able to establish that organic methods were used continuously during at least the last third of the gestation period. This means that your swine must be fed entirely on pasture, forage, crops, and feeds that are organic during this period. In addition, organic pigs must live in a habitat that approaches a "natural" setting, allowing them to develop as a part of nature.

Many people believe that organic foods are healthier for human consumption than those foods produced by standard methods, though the science of this assumption remains controversial. Organic fruits and vegetables tend to be smaller and more irregular in shape and color than their commercial counterparts, though if you have ever eaten them, it is hard to deny that they seem more flavorful and satisfying. Animals raised by organic methods tend to be higher in nutrients and lower in nutritionally undesirable compounds, such as heavy metals, fungi, pesticides, and glycol-alkaloids, which are naturally occurring toxins found in some feeds.

An organic livestock operation must also handle animal waste very carefully. Manure cannot be allowed to contaminate water, soil, or crops, but it should be used to maximize reclamation and recycling of nutrients. You will also have to monitor what you feed your pigs if you wish to adhere to organic standards. Alfalfa, barley, clover, and corn are all examples of crops you can grow for hog feed. Organic farmers can also purchase organic feed for their pigs if they cannot grow their own crops. When swine are pastured and permitted to root, they turn over the soil and apply manure at the same time, preparing your meadows to receive a feed crop.

Some small organic farmers keep a pig fed on table scraps on the farm just for the manure, which makes excellent fertilizer. That kind of enrichment can have a beneficial effect on your soil, providing a lush, green lawn or a vegetable garden to supply your table with the best produce you have ever eaten. If nothing else, you may want to supplement your pig's feed with a patch of alfalfa or several rows of corn. Any crops you grow yourself to feed your pigs will reduce the feed you need to purchase, and you will know, from seed to harvest, what you feed your pigs and whether it follows organic principles. It is definitely something to consider if you are inclined to be green.

Many conventional methods may not be used to maintain the health of your herd if it is to be certified as organic. For example, using preventive antibiotics is prohibited, and although a veterinarian may prescribe them as treatment for an illness when appropriate, animals that have received such treatment may no longer be marketed as organic. In essence, the central idea of organic pork production is that swine health can regulate itself, if you have provided the animals with a suitable environment that minimizes stress and promotes health.

Pasture Raising

An old approach to raising pigs that is becoming popular again is to raise them in the pasture. Although pigs are not thought of as pasture animals in the same way that horses or cows are, they can make good use of a field of grain crops and some alfalfa, supplemented with some corn feed. Sows can also farrow outdoors in farrowing huts and raise their litters, though they still need plenty of human attention. Many of the older, rarer breeds are known for being good for raising outdoors. The Gloucester Old Spots has long been known as the "orchard pig" because it used to be kept in orchards and fed off fallen apples and fruit from trees. Another old breed, the Tamworth, can do very well in a pasture situation.

Heritage breeds of swine are breeds that have a long history, sometimes going back several hundred years. They have been used to create today's modern breeds of swine. Heritage breeds do not grow to be as large as the hogs raised by commercial breeders in confinement facilities; however, there is a strong market for heritage meats. They have stronger and more varied flavors than commercially raised pork, and they are favored by many chefs and foodies who are willing to pay more for heirloom pork. Many heritage breeds are ideal for being raised in the pasture and do well by foraging on nuts and fruits if you have these available on your property.

Texas Tech University studies ways to make pasture raising pigs both economically and environmentally sound on a large scale for farmers. In their Sustainable Pork Program, Texas Tech University began studying intensive outdoor pig production in 1993. In 1998, they created a research farm to examine ways to make pork production "animal, environment, worker, and community friendly." Texas Tech uses a paddock system that allows 100 acres for every 300 sows. The paddocks are arranged in 12-acre lots that radiate out from a central hub where the pigs can be handled and watched. The various paddocks are arranged for breeding, gestation, farrowing, and pasturing. The university's researchers study and evaluate all aspects of the program, such as production costs, the pigs' behavior, the effects on the environment, the quality of the pork produced, and microbe levels in the soil. So far, the researchers have found an improvement in pig health, a better work environment for workers, reduced odor, reduced microbial activity, fewer regulatory problems, and lower costs for start-up and operations. In addition, they found that it cost $23.20 to raise a pig in this kind of intensive outdoor pig production system, compared to $31 in a typical confinement system (in 1995). There was a net profit of $10.39 in raising a pig in the outdoor system in this study.

Finding a Breeder

Now that you have had a chance to learn something about the different approaches to pig farming, it is time to start thinking about your plan. If you are just starting out, both organic and commercial methods might be a little ambitious at the outset. Instead, you will probably want to start off with a small modified-confinement system because this method is best suited to the needs of a beginner.

In this section, you will learn all about purchasing your first young, recently weaned pigs, or shoats. But, it is important to have their living area established before you bring them home, and you will need to conduct a little research to find the best breeders in your area. If you know anyone in the business, you have a tremendous advantage, but you can learn a good deal by talking with local feed store owners and veterinarians. Check your newspaper's classified ads, Yellow Pages, or Craigslist (**www.craigslist.org**), but be sure you have also consulted with your region's pork producers' associations or breed associations, such as the American Berkshire Association, the United Duroc Swine Registry, or the American Livestock Breeds Conservancy. *See the Appendix for a more comprehensive listing of breed registries.* Breeders that join such associations tend to be more concerned with their reputations and thus work hard to ensure their stock is healthy and well bred. Finally, you need to check into your community's zoning laws to ensure they permit pig ownership.

If possible, never buy your pigs from a stockyard, auction, or other third-party seller. Buy directly from the farm of origin if you can because the more livestock is transported, the more chances there are for something to go wrong. Transport to an auction means contact with pigs from other farms, which means a greater likelihood of disease and injury.

When you buy directly from the breeder, you have a better opportunity to assess the conditions of the place where your new pigs originate. You can learn much just by looking. If the farm is in disrepair and not well maintained, you can be sure that the animals on it have not fared any better. Dirt and manure are a normal part of any farm, but a good farmer keeps the farm in order so it functions well. Animal enclosures should not have significant manure buildup, and bedding should appear fresh and clean.

Choosing Your Pigs

Once you have settled on a breeder, you can begin to think about making a purchase. One of the best beginners' approaches is to purchase a couple of shoats in the spring and fatten them for fall hams, keeping one for the table and selling the other, if you so choose. There is some division about whether it is better to work with barrows — young, castrated male pigs — or gilts —young female pigs — though each has its advantages. Barrows tend to gain weight faster than gilts, but gilts can accept more nutrient-rich feed and produce leaner meat. In general, either will do, and your decision should really be based on the best-looking pigs available.

Early in the spring, contact the farms and breeders in your area to find out when they are farrowing — that is, when their sows will be giving birth. You will want to buy your feeder pigs when they are in the 6- to 8-week-old range. This is a good time to buy your pigs because their prices are at their lowest at this age, and they are easy to handle. As the piglets get older and the farmer invests more in feeding them, their cost will increase. Sometimes, it is possible to pick them out before they are weaned and return for them later. If this is possible, you may want to do so. Good breeders castrate their young male pigs as a matter of routine so this should be done, and the animals should be well healed before you bring any home.

When you go look at the pigs, the farmer will most likely have all the young shoats that are for sale in a single pen. Take some time to look them over. Healthy pigs will move far away from visitors and then gradually approach as they grow more accustomed to the newcomers and their natural curiosity takes over. Pigs that make no movement when strangers approach should be viewed with suspicion; such listless behavior is often a sign of poor health or injury.

In general, crossbred or hybrid pigs tend to be more vigorous than purebreds, although only when the crossbreeding is part of a planned breeding program. Ask the farmer about his or her breeding approach; most will be happy to talk about their ideas about breeding because it says much about who they are as farmers. If possible, you want to take a look at the parent stock as well to see what kind of animals you are buying. Look for sows with smooth shoulders and skin, firm jowls, and a neck neither too long nor too short. Look at the pigs' heads, which should appear proportional to their bodies — a large head with a small body is a bad sign, suggesting that the pig has not grown properly.

Look at all the pigs in the group. In your mind, separate all the largest animals first because you want to take home the biggest in the lot. Discount any that do not seem lively or are notably lethargic. Watch out for animals that limp noticeably, have diarrhea, or keep themselves separated from the other pigs. Once you have the field narrowed down to half a dozen or so, tell the farmer which ones you want to look at more closely. The farmer should help you by catching these pigs, and you want to observe them during this process. A pig that does not try to get away, squealing in protest, is suspect. Closer examination should reveal a pig with a nice, glossy haircoat and a vivid skin color — with white pigs, this should be bright pink. Look behind the ears for parasites, look in the eyes for discharge, and examine the feet for soreness. Sometimes a farmer lays down extra straw to hide tenderness in the feet, but you can detect this in the final examination.

A pig's snout should be straight, not twisted or malformed, and there should not be any excessive runniness from its nostrils. Watch for sniffling and shaking, signs of respiratory problems and sickness. Feel around the jaws for lumpy nodes or pustules, which can be a sign of abscesses in the jaws — these can be treated, but you do not want to buy a pig if you notice these problems. Next, look at the pig's eyes, which should be bright, free of discharge, and widely spaced, which many farmers feel is a sign the young pig will grow into them. Similarly, a pig with thick bones, loose joints, and large extremities — such as ears, tail, and feet — is thought to have more growth potential. Have you ever seen the puppies of a large dog breed like Great Danes? The larger their ears and feet seem in relation to the rest of their body, the larger they grow — it is similar with young pigs.

According to the USDA, the average prices of feeder pigs in the early spring of 2010 were approximately $40 to $50 per head in the 10-pound range and $65 to $75 in the 40-pound range. These prices refer to pigs bought in larger quantities so you can expect to pay a little more for individual animals. Again, local knowledge will be invaluable to you in this area; if you know someone in the business, you will be at an advantage. Feed stores, custom slaughterhouses, and vets can provide some assistance.

Beyond simple weight and age considerations, the relationship between weight and age is an important sign of an animal's potential for growth.

Feeder pigs should be around 35 to 45 pounds at 8 weeks, and 60 pounds at 12 weeks. There is some variation by breed and gender, but these figures are meant to be guidelines, not fixed rules. Just make sure you do not buy a 12-week-old pig that only weighs 20 pounds, unless it is a miniature breed and you want a small one.

The USDA also has a grading system for feeder pigs based on their "logical slaughter potential" and "thriftiness." These are technical terms that refer to a pig's growth potential and feed-to-gain efficiency. There are six categories (Grade 1 being the best), but you should not purchase a pig of worse than Grade 1 or 2.

- Grade 1 pigs are long bodied, with thick muscling throughout and full hams and shoulders that are thicker than the rounded back.

- Grade 2 pigs appear similar to Grade 1 pigs, but their muscling is slightly less thick, and they are slightly shorter bodied. It is unlikely that, as a beginner, you would be able to tell the difference between a Grade 2 and a Grade 1 pig, but both make very good starters.

- Grade 3 starter pigs seem short in relation to the size of their heads, and the muscling of their bodies appears thin over flat backs, with narrow shoulders and hams.

- Grade 4 pigs are decidedly short, and their thinly muscled bodies appear flat and thin, particularly in their lower regions.

- Utility Grade pigs are very thinly muscled throughout the bodies and have an unkempt appearance with tapered legs and thin hams and shoulders.

ⓢ Cull Grade pigs seem very weak as a result of disease or poor care, and they can only make a normal market weight after an extremely difficult and costly feeding regimen.

Choosing Your Breed

Breed selection is an important part of the decision-making process for new pig owners, whether you intend to keep them for pets, for breeding, or for the table or market. In this section, you will learn about common breeds, which should help you form some idea of the type of pig you want. Your geographic location is very important because different breeds do better in different climates. Darker pigs are common in sunnier southern regions because their pigmentation helps protect them from sunburn. If you live in a colder state, you may wish to choose a breed with a shaggier coat that can help them stay warm in winter months.

A key factor in making this decision is which breeds are available in your area, but it is also important to consider what you plan to do with your pigs. If you plan to establish a breeding program, you want to buy the best stock you can find. Purebred pigs are pedigreed, carefully regulated, and expensive. If you just plan to fatten a pig or two for the table, crossbred pigs will do nicely, as long as they are produced by a knowledgeable, responsible breeder with a purposeful breeding plan.

Pet breeds

"I am fond of pigs. Dogs look up to us. Cats look down on us. Pigs treat us as equals."
Winston Churchill

There are a number of breeds that are more suited to pet ownership than the larger meat breeds. Many of these pigs are still raised as food breeds, but their smaller size makes them better as pets. Because of their great similarity to humans, both in physical attributes and in their social behavior, they can fit right into a human household under the right circumstances. It is important to consider, though, whether your household will fit right in with them. Are there members of your family who would be incompatible with pigs? Do you have other pets that might behave aggressively toward them?

Because pigs are such social animals, they can form strong bonds with people and other animals. You should realize that this also means they often require companionship so you will need to consider whether you can provide them the type of home they need. Do you have the space needed? Can you devote the appropriate amount of time to your pet pig?

Miniature pigs can be either midgets, which means they are much smaller, but proportionally identical to their larger counterparts, or dwarfs, which means they are smaller and proportionally different, such as potbellies.

African Pygmy/Guinea Hog

The African Pygmy, or Guinea Hog, is a small black pig descended from larger red hogs thought to have been imported to the Americas from Africa aboard slave ships. One reason African Pygmies make good pets is that they have long life spans, sometimes reaching 25 years. An African Pygmy is only 40 to 60 pounds, making this a very manageable breed. They have kinked tails, straight backs, and medium pricked ears. Their shiny black coats are hairy rather than bristly, which makes for better cuddling. Not a potbellied breed, African Pygmies are grazers, preferring lush, green grasses. This is a very friendly, adaptable breed that can be a cute, cuddly companion for many years.

Juliani/Painted Miniature

Painted miniature pigs were imported to the United States from Europe. They range from 15 to 60 pounds and have a small potbelly, a slightly swayed back, and proportionally longer legs than true potbellies. They can be black, red, white, silver, or a mixture of these colors. They are among the friendliest, most playful miniature pigs and are considered extremely gentle. Also, they are so small that they integrate more easily into an average home — some are even kept in urban apartments.

CASE STUDY

Gary and Shelly Farris
Rocky Mountain Kunekunes and Idaho
Pasture Pigs
PO Box 765
Rigby, Idaho 83442
Kunekunes@q.com
www.rockymountainkunekunes.com
208-745-7978

Gary and Shelly Farris own Rocky Mountain Kunekunes and Idaho Pasture Pigs in Rigby, Idaho. According to Shelly Farris, they raise a little of everything on their farm, but their primary focus is pigs. She and her husband, Gary, have been around pigs all their lives, and they started raising them together in 2005. Gary wanted a smaller pig to keep on the farm, something friendly and pasture based. He researched pig breeds for months until he found the Kunekunes and fell in love. The Kunekunes are a small pig from New Zealand. They are extremely friendly, and Shelly says they are a joy to have around. "They follow us around like faithful dogs and always want belly rubs and to be petted."

Shelly Farris says that while she loves pigs, she wanted meat for the freezer. They compromised and got a Duroc, too. That was the start. Soon they added more Kunekunes and Berkshires to the mix. "The contrast between our 'big' pigs and our 'little' pigs is amazing. We have two worlds really because they are so different."

Farris says that the pigs are pasture based in the summer, and they feed grain and alfalfa that they grind and mix in the winter. "We raise our animals as naturally as possible. We do not use hormones and only use antibiotics when necessary. We have studied the history of pigs and how they were raised a 100 years ago. We have studied grasses and legumes and planted our pastures accordingly. We incorporate AI (artificial insemination) on our farm and also use boars for natural breeding. We do not use gestation or farrowing crates but choose to farrow in barns, huts, and **loafing sheds** (a simple lean-to shed so pigs can get out of the rain and weather). Farrowing in the winter in our climate is doable but challenging. It requires a lot of work and diligence to keep the piglets warm."

She said that they have a relatively small herd of 25 to 30 pigs at all times. The number of pigs goes up when they farrow new litters, and the number of pigs drops when the animals are sold and move on. She says they keep mostly a closed herd so they will not bring in diseases or new parasites, and only introduce a few new animals to the herd. Instead they introduce new bloodlines to the "big" pig herd by using artificial insemination.

Farris said, "We want to bring back the pig to family farms across the country. Through selective breeding, we are developing the perfect pasture pig. Snouts on pigs have been lengthened over time, and we are breeding for shorter snouts. We have found that the Kunekunes do not root and have a really short, upturned snout. We selected other breeds like the Berkshire and Duroc that also have relatively short snouts and have been breeding for that trait, as well as others."

Farris says that using pasture rotation has been the best method of keeping pigs for them. "In the summer the pigs are out on pasture. We have found that pasture rotation is the best method for us. Using other farm animals like cows, goats, sheep, horses, turkeys, and chickens in the rotation is beneficial to pasture management. The horses like the young tender grass, the cows like the taller grasses, sheep and goats eat grasses that others don't, and the pigs like a medium grass, especially clover, alfalfa, and orchard grass. We have more animals than we have ever had on our pastures, and the pastures are in the better shape than they have ever been. We now have so much grass that we actually need more animals to eat it all off. Pigs on pasture spread manure on the pasture as they graze. It is the consistency of horse manure and does not smell."

Even in winter the pigs still do well outside. "Our winters can be really harsh as temperatures can get 20 to 30 degrees below zero and much colder than that with wind chill. All of our pigs are outdoors, but they all have shelter in the form of huts, barns, and loafing sheds. These we fill with straw. We keep piglets in the barn with heat lamps. We farrow in barns and have farrowed in the loafing shed. We feed heavily in the winter as they need more feed to stay warm."

For butchering, Farris prefers to have a mobile butcher come to the farm. "We used to haul the pigs to the slaughter house but found that it was really stressful on the pigs and ourselves. We now have a mobile butcher that comes to our farm. The pigs are in their natural environment, usually eating or talking and do not know what hits them. There is no stress. We raise our animals in a caring environment and to us this is the most humane way to end that life. We always taught our kids that we are going to raise this animal, care for it, love it, and feed it and in return it is going to feed us."

Kunekune

This fine pig is a more recent addition to the family of pet pigs in the United States. Imported from New Zealand, their name comes from the Maori word "kune," which means fat and round. Their most distinctive feature is probably their wattles — tassels hanging from their lower jaw — known as pire pire in the Maori language. Their ears may be erect or semi-lopped, and they come in a wide range of colors: black and white, red, white, brown, and gold. A midsized pet pig, they range from 90 to 120 pounds, have attractive pug noses, and may have spotted or calico coats.

Kunekunes are efficient grazers so during the summer months, they may sustain themselves just by keeping your lawn well trimmed. In addition, they do not root so they will not spoil the appearance of your grass.

Vietnamese Potbellied

Developed from the "I" breed of Vietnam in the 1950s, potbellied pigs are among the most common pet breeds in the United States. Keith Connell, who saw their potential as zoo animals, first brought them into the United States through Canada in 1985. As the name implies, the potbellied pig has an exaggerated potbelly with a swayed back, erect ears, extremely short legs, and a straight tail. Potbellies have very wrinkled faces and a short snout, giving them a somewhat comical expression. When purchasing a potbellied pig, try to be sure that the short snout is not too exaggerated because this can cause respiratory problems. Potbellies are normally black, white, or a piebald pattern of black and white.

They are often found in zoos because they have an appearance many people find appealing, and they are friendly and even-tempered. In fact, because of their docile disposition, they are a mainstay of the petting zoo; not only do children love them, but they also seem to love children. This friendly manner and sociable behavior has made them the most popular miniature breed in the world.

Potbelly pigs may be very small, as little as 25 pounds, though most are between 100 and 250 pounds. If you are thinking about buying a potbelly, this is an important consideration because although the young pig is a tiny animal, it may grow to the size of a large adult human. If you are looking for a petite pig, it is a good idea to look over the breeder's adult animals. Too many have bought a tiny pig only to realize that the adult pig is more than they can handle. This is a growing problem in the United States, and as a result, it is increasingly common to find pigs in animal shelters.

Yucatan/Mexican Hairless

Originating in Central and Latin America, the Yucatan is a very gentle breed of pig, ranging from slate gray to black in color. Yucatans have straight backs, short snouts, and medium-sized ears that are erect. Their skin and body systems are probably the most similar to humans, which has made them the most common pig used in laboratory testing. Yucatans can reach 200 pounds, though some considerably smaller strains are bred.

Breeds for food production

Although all pig breeds are edible, most pig owners prefer to draw a relatively firm line between pet breeds and food breeds. Not only is this important from an emotional standpoint, but food breeds also tend to be larger and have other characteristics that are more desirable for farmers, including their breeding qualities, docility, efficiency of weight gain, and, of course, meat quality. The food breeds raised for meat are divided into three categories: meat types, bacon types, and lard types. Meat hogs tend to have very large frames with leaner bodies that provide a fine grade of muscle. Bacon hogs, on the other hand, tend to have very long bodies that provide larger sides of bacon with plenty of lean meat. The third type of hog, the lard hog, has more or less disappeared in recent years, as the demand for leaner pork has grown and the use of lard in cooking has decreased.

Eight breeds of pig provide most of the pork produced in North America, with commercial breeders primarily concentrating on complicated crosses of three breeds: the Duroc, Hampshire, and the Yorkshire. Crosses of these breeds in succeeding generations, along with crosses to the Landrace, the large English white, and a few other breeds, are used to produce large litters, the most efficient feed-to-gain ratios, and other measures that are important for commercial pork producers.

On the other hand, small farmers are better off focusing on raising purebred pigs, or simpler crosses. Instead of using complicated breeding schemes to achieve super pigs, or pigs that have been created from multiple hybrid crosses like those used by commercial pork producers, it is better to focus on one breed that produces large litters and has few losses before weaning. Look for breeds with good mothering traits that produce plenty of milk for their young. Consider hardier breeds that will thrive in a modified confinement setup instead of living in a confinement system. You may also wish to consider some of the breeds that have slower weight gain, especially if you have a local restaurant market with owners interested in trying more flavorful pork.

Berkshire

Berkshires were presumably "discovered" by Oliver Cromwell's troops when they were stationed at Reading, England, during the English Civil War in the 17th century. Berkshires have such flavorful meat that they fast became the most popular breed among England's upper classes; indeed, they kept a herd at Windsor Castle within sight of the royal residence. Maintained as a distinct breed, they are widely considered "England's oldest pig."

When imported Berkshires reached the United States in 1823, they had a sandy or reddish coloring. But, they were quickly crossbred with other breeds, leading to the color pattern seen today: black with white feet, snout, and tail, identical to the Poland China pig. The Berkshire is a smaller hog, though, with boars averaging 500 to 750 pounds and sows 450 to 650 pounds. In 1875, a group of Illinois breeders and importers formed the American Berkshire Association to ensure the continuation and preservation of the breed and its conformation standards, including short, erect ears and a medium-dished face with very deep sides.

Berkshires offer a good growth rate and fair reproductive efficiency, though their litter sizes tend to be relatively small. However, the quality of the Berkshire's meat has made it a favorite of gourmet cuisine, with its exceptional marbling and flavor. The meat is also prized in Japan, where they are bred in the Kagoshima Prefecture.

Chester White

The Chester White is an American hog breed that originated in Pennsylvania in the early 19th century, based on a cross of the English Chester, Lincolnshire, and Yorkshire breeds. Although only a medium-sized hog, the Chester White is popular with breeders and packers because of its high quality of muscle tissue and because its lighter color has an appealing appearance to consumers. In addition, the Chester White has a high degree of **cutability**, meaning that a larger percentage of its mass translates into marketable cuts of meat.

The biggest advantage of the Chester White lies in its exceptional breeding abilities. A Chester sow breeds back very quickly; that is to say, once she has farrowed, she may breed again more quickly than is common in other breeds. Chester White sows are known to farrow as many as three litters a year, and these litters often contain ten or more pigs that reach market size. In addition, the sow carries these strengths when bred with other types of pigs, making them a popular choice in crossbreeding programs.

The Chester White has a medium frame, a slightly dished face, medium-sized lop ears, and a thick, white coat. In addition, the Chester White is an extremely sound animal that can maintain its health in different conditions so smaller farmers with simpler facilities and outdoor pastures often find them a good choice.

Duroc

Durocs are a red hog that range from a tawny golden color to a deep mahogany red. Durocs have large lop ears that hang over their eyes, a curly tail, and a slightly dished face. The average boar weighs about 900 pounds, and the average sow reaches around 750 pounds. They are considered a good meat breed because of the high quality of their muscle tissue, large bodies, and relatively low fat content.

Considered a Northeastern breed, there is some dispute as to these hogs' origin. Some claim they descended from the pigs brought to the New World by de Soto and Columbus, while others believe their ancestors were the African Guineas that may have come along with the slave ships. As a distinct breed, they originated in the early 19th century with a farmer named Isaac Frink of Saratoga County, New York. As the story goes, Frink was visiting his neighbor Harry Kelsey's farm when he took a liking to some reddish hogs. He purchased a few of them to start his own herd, and because the breed was unnamed, he decided to call them Durocs after Kelsey's famous Thoroughbred stallion. Later, they were crossed with Jersey reds, producing a hog that has developed into one of today's most popular breeds.

The Duroc is among the most common breeds found in the United States. The boars tend to be aggressive, and they are often used in crossbreeding programs, especially with Hampshires or Yorkshires. The sows can produce large litters, and the young pigs gain weight faster than almost any other breed. This, along with their extreme hardiness, accounts for their popularity.

Hampshire

Hampshires are an English breed, though the breed that is well known today was developed in Kentucky. With erect ears, they are mostly black, except for a white band across their shoulders and forelegs; because of this distinctive marking, they are also known as saddlebacks.

Hampshires are somewhat smaller than other meat hogs, with sows reaching about 650 pounds and boars reaching 800 pounds. They are very low in lard, with high-quality meat and a large loin eye area. In addition to their high-meat quality, Hampshire sows are exceptional mothers who remain fertile longer than most. Their breeding potential and high quality meat make them a popular pig. According to the National Swine Registry, Hampshires are the fourth most recorded breed in the United States.

Hereford

The Hereford is the most American of hogs because they are not bred in any measurable quantity anywhere else. Their fans consider them the best-looking pigs, and they certainly are distinctive, with their flashy red and white coloration like Hereford cattle. The Hereford has a slightly dished face, medium-sized lop ears, and a deep red back with white trim around its legs, head, and tail. John Schulte of Norway, Iowa, originated the breed in 1920, based on a cross between the Chester White, Duroc, and Poland China breeds.

With boars averaging 800 pounds and sows averaging 600 pounds, Herefords do not grow as large as other meat hogs, but they are quite popular for other characteristics. Herefords develop very rapidly, reaching maturity at 200 to 250 pounds in just five or six months, and they do so on less feed. Herefords are adaptable to a variety of different climates, and because of their quiet demeanors, they are a popular choice for youngsters engaging in 4-H Club and FFA projects. The sows make excellent mothers, producing and weaning large litters. This, combined with their efficient feed use and rapid maturation rates, also mean that Herefords can be very profitable.

Poland China

Strangely, Poland China hogs are neither Polish nor Chinese. In actuality, they originate in the Butler County/Miami Valley region of Southeastern Ohio. The Poland China is a large, black hog with white "points" — face, feet, and tail — and lop ears. They tend to be long bodied, lean, and muscular, making them an ideal meat type, with average boars of about 900 pounds at maturity and sows of about 800 pounds. Big Bill, the largest hog ever recorded, was a Poland China of 2,552 pounds owned by Elias Buford Butler of Jackson, Tennessee, in the early 1930s.

Like Durocs, Poland Chinas are very hardy animals that feed very well. In addition, they are exceptional breeders that are well suited to transportation because of their quiet dispositions.

Spotted Poland China

Spotteds are actually so closely related to Poland Chinas that they could almost be considered the same breed — the spotted looks just like a Poland China but with spots. But, spotted breeders consider theirs to be a better pig, and they have been organized under their own breeding association. They look very similar to Poland Chinas, although as their name implies they have a spotted coloration, either white with black spots or black with white spots, which they inherited from the Gloucestershire Old Spots side of their family.

Most feel that spotteds offer only a moderate meat quality, although the sows are known for their exceptional mothering ability. Not only do they produce a good quantity of milk for their young, but they also give birth to very large litters — one of the largest of the colored breeds. Like Poland Chinas, they are very good feeders, maturing early and growing very rapidly.

Landrace

Landrace hogs originated in Denmark, which jealously guarded exportation of these hogs for centuries. However, in the 20th century, importing these fine animals became possible. They have a soft, white coat with pink skin, long, drooping ears, and flat backs. Although their legs tend to be short, they have a long, lean body that makes them an ideal bacon type. In addition, their long bodies have 16 or 17 pairs of ribs, which results in more cuts of meat. The typical pig has 14 pairs of ribs.

Landrace pigs tend to be quite docile, and they grow very rapidly. One of their most desirable characteristics is that Landrace sows produce unusually large litters, as well as a great deal of milk with which to support their piglets. As a result, Landrace sows are often used in crossbreeding programs, particularly with Durocs.

Yorkshire

Like the Landrace, the Yorkshire pig is a bacon type with white hair and pink skin. It has a dished face with erect ears and a long, lean frame that supplies ample, high-quality bacon. They were brought into the United States in the early 1800s, though it was not until the 1950s to 1970s that the breed really flourished. Yorkshires are comparatively small, with mature boars averaging 600 to 800 pounds, and they tend to grow slowly. Nevertheless, it is a popular breed commonly found in commercial pig farms across the United States.

Yorkshires breed well, with large litters and sows producing plenty of milk. Yorkshires are also commonly seen in the media — if you can think of a famous pig from TV or film, it was probably a Yorkshire. For example,

Arnold Ziffel, who often upstaged Eva Gabor on TV's *Green Acres*, was a Yorkshire, along with Babe from the 1995 film of the same name.

Ossabaw Island

Feral Ossabaws inhabit the island in Georgia from which they draw their name. In the 1500s, Spanish explorers often left small herds on islands in the Americas to establish future sources of food, and the pigs of Ossabaw Island are thought to be descended from one of those herds. Although they may be seen as a pet breed by some, Ossabaws are prized for their dark, unusually textured meat, as it resembles that of the black Iberian pig.

Living in an isolated island environment has had some interesting effects on the breed. One consequence of their isolated existence in a sparse environment is an extremely high level of intelligence, which they require in order to exploit every possible food source. In addition, Ossabaws carry a "thrifty gene" that permits them to store fat effectively. The consequence is that domestic Ossabaws, which have ready access to ample food supplies, often develop a form of diabetes. Because of the breed's problems with diabetes, the breed is rarely ever crossbred with other pigs. Ossabaws should be between 14 to 20 inches tall and 25 to 90 pounds.

Ossabaws come in a wide range of colors, but unless crossed with pigs of other breeds, they never develop stripes. In general, they have solid or spotted coats, sometimes resembling a calico pattern. Ossabaws may be red, gray, blue, and even white, although this is quite rare.

Ossabaws today are critically endangered. They are currently found in a few zoos on the mainland but, in general, the pigs are not allowed to be removed from Ossabaw Island because they are at risk of carrying porcine vesicular stomatitis and because pseudorabies is found on the island. There are fears that the pigs on the island could transmit these diseases to pigs on the mainland.

Pigs for composting

Although it would be unusual to get pigs for the sole purpose of using them to help with your composting, if you are a homesteader or if you have a small farm, you may wish to consider adding two or three pigs to your livestock herd. Pigs are excellent at rooting and turning soil.

In order to encourage pigs to help with your composting, you can set up a small fence capable of holding two or three pigs around your large compost heap. Electric netting or Pig Quik Fence would work well. *See Chapter 5 for portable fencing options for pigs.* Many people place a layer of straw over the compost so the pigs can work it into the compost as they root. The pigs will turn over the compost thoroughly trying to find anything good to eat in it.

It is a good idea to have other uses for these pigs in addition to the composting, as the composting will not keep them busy for long. If you are not interested in raising pigs for this purpose, you may wish to consider borrowing pigs for a day or two from a neighbor who raises pigs. You will need to have a very large compost heap to keep the pigs interested and busy. Even if you add material to the compost heap every day, you would need to supplement the pig's diet with more nutritious food.

You could use any kind of pigs for composting, though it may become difficult to keep larger pigs inside a small enclosure.

Summary

In this chapter, you have considered different ways to raise your pigs and different approaches to confinement. Large commercial producers commonly use indoor confinement, while small farmers opt for modified confinement with a sleeping area and a pen. Pasture raising pigs is also a popular option for raising pigs. You have also considered raising pigs organically. *There will be more discussion of organic methods, especially organic feed, in Chapter 6.*

You have also considered a few of the many breeds of pigs you may consider raising. Most of the more popular breeds are easy to obtain, but if you wish to raise a particular kind of purebred pig, you may need to look farther away from home to find a breeder. You have examined some places where you may find pigs for sale, such as through local farmers, your local feed store, Craigslist, and through breed associations. *There is an expanded listing of associations and contacts in the Appendix.*

In the next chapter, you will consider caring for young pigs and the needs of piglets. Whether you are raising piglets farrowed by your own sow or you have purchased young shoats, you will need to know how to care for the young pigs. You will also consider what to look for when choosing young pigs. Your pigs will depend on you for their care so you will need to know what to do right from the start.

CHAPTER 3:

Caring for Young Pigs

As you have already seen, there are many breeds of pig from which to choose before you get started. Before you think about bringing any pigs home with you or having any sows farrow, you will need to make sure you have your property set up and ready to receive your pigs.

You should have a pen ready with a sleeping area or a pasture with a good hut. *See Chapter 5 for more information about pens and sleeping areas, as well as waterers and troughs.* You should have your feed ready, which will be discussed in Chapter 6. Once you have everything ready, it is time to bring home the pigs. This chapter will discuss caring for young pigs, whether you purchase young shoats or they are born to your sow.

Educating yourself about the possible pitfalls and problems in rearing young pigs is the best way to ensure success. The first two to three weeks are the most sensitive in pig care. It is imperative you ensure they take in the proper amounts of nutrition and the environment is clean, dry, and comfortable.

Caring for pigs becomes less of a chore as they grow and mature. They love to interact with humans as much as their littermates. The curious nature of a pig reveals itself completely with the antics they display. The constant variation of frolicking one moment and standing as still as a statue the next when hearing unfamiliar noises will have you rolling with laughter.

With proper attention to detail, any pig farmer can enjoy the success of raising happy and healthy litters. Recognizing the signs of healthy newborn care will make purchasing and raising pigs an enjoyable experience. Generally speaking, a happy pig is usually a healthy pig.

Characteristics of Newborns

Playful and curious tend to be the overwhelmingly common descriptions regarding young pigs. Though they start very small in size, approximately 3 pounds and 8 inches long, their growth rate is phenomenal. To kick the growth spurt into full drive, it requires plenty of eating and sleeping.

When purchasing a shoat, it is critical that they have been fed colostrum in some form in the first 24 hours to make sure they have received a good start in life. If there is doubt about this, it is best to pass on purchasing that shoat. These initial feedings pump in the necessary ingredients for a strong immune system, passed through the colostrum from the sow. Of course, the most natural way for young pigs to receive colostrum and the antibodies it contains is by nursing from their mother. Continued regular nursing produces a healthy pig of acceptable weight and size.

Runts of the litter are pigs that have been deprived of feedings at some level. This translates into a pig that may be more prone to illness and will never regain the growth opportunity lost. Healthy pigs are more than willing and able to nurse from the moment of birth.

The tiny stature of a newborn pig leaves them prone to injury. The most common infant fatalities are from accidental crushing by the sow. Rolling over on the newborn pigs will cause significant problems and possibly death. Maintaining an area a few feet away, complete with a heat lamp, will prevent most injuries of this type. Pig rails and creeps, or areas set aside by rails or bars where the sow cannot follow the piglets, are a common feature of many farrowing spaces in order to allow the newborns to roll away from their dam and avoid being crushed.

Even though pigs are born with their eyes open, they spend a majority of their time in the first week or two with their eyes shut. Sleeping is a necessary and favored activity of newborns. The body is using so much energy to grow that there is little left for other activity. After three weeks, they will begin exploring their environment. Pigs that isolate themselves and sleep more than the rest of the litter might be experiencing health problems.

Special Needs of Piglets

Simply put, pigs are babies, and babies of any species require extra care and sensitive treatment. You can breathe a little easier if you can get a pig past the weaning phase because most fatalities occur in the first few weeks of life. Each day is another bridge crossed and another milestone of success.

Do not be afraid to jump in the pen with them once they have weaned. The more you familiarize yourself with healthy young pigs, the more you will develop a keen eye to spot problems early on. They will squeal with delight to interact with you.

On a serious note, things can take a turn for the worse quickly with piglets. There are some serious diseases and illnesses out there, and you must be diligent about watching for signs and symptoms. Early and aggressive treatment could very well be the thing that saves their life. The following are things that need to be highly monitored or given special consideration when raising young pigs. There are warning signs when newborn pigs are in trouble, but there is often little time to act on these. Spending quality time with the litter will help you clue in on changes and problems before they reach a critical point.

Colostrum

The intake of colostrum is possibly the most vital way to ensure a pig's survival, second only to taking in that first breath. Colostrum is the sow's first milk, produced 12 to 24 hours after giving birth. The inability of the pig to drink in this nutrient- and antibody-rich substance could spell disaster. Pigs known to not intake colostrum show rapidly declining health. Survivors are plagued by low lifetime weight and height gains. Lack of immunity to the most basic of germs and disease means the pig is susceptible to any and all illness.

Extra colostrum can be collected and stored in frozen form. Storing it as cubes tends to be the easiest, most convenient fashion. It must be thawed before use. Boiling or microwaving colostrum will destroy the nutrients and antibodies, thus rendering it useless. If you heat colostrum, you should simply warm it up to body temperature for the young piglets (101.5-102.5 F).

You can gather or purchase alternative forms of colostrum. The best substitutes seem to be from cows and goats. The colostrum these animals produce is not quite as nutrient rich, but it will do when no other source is available. You can also purchase colostrum mix from feed stores in an emergency.

Temperature

Newborns of any species enter the world with a total dependence on care. The provision of food, shelter, and a clean environment falls on the sow, as well as the pig owner. One often overlooked ingredient for raising healthy pigs is temperature.

Temperature stress is remarkably easy to spot in young pigs but too often ignored. Piglets must be maintained in a dry, clean, comfortable environment in order to thrive. Newborns that are too hot or too cold are equally harmed. Of course, sows may be bred at any time when they come in season and have a gestation period of 114 days, which means that litters may arrive year-round. It is normal to rebreed three or four days following weaning when the sow comes in season again. This is considered an optimum time for breeding. Sows have two to three litters of piglets per year.

Pigs that are cold will pile on top of one another. A certain amount of piling is normal, but a constant bid to seek the warmth of others is a signal that the air temperature is probably too chilly. The hair will be raised in an effort to insulate the skin from cold drafts. Extreme cold leads to poor blood circulation, respiratory difficulties, and death.

Overheated pigs will separate out from one another. This is highly unusual behavior with newborn pigs and is a red flag that there is a temperature problem. Panting and deep breathing are both indicators that they are too hot. Touch the pigs when unsure of their comfort level. Dry, hot skin means the temperature needs to be brought down. Cold, clammy skin means that the pigs are too cold and may be suffering from exposure.

Dehydration

There is a world of difference between a newborn pig and the full-grown pig it will become. Though ending up hearty animals, they begin as fragile babies that need monitoring for environmental and dietary sensitivity. Needing to suckle up to 16 times per day, a few missed meals will quickly trigger dehydration. This leads to serious weight loss and death in the most serious cases.

Diarrhea in young pigs, also known as scours, has many causes. The more typical causes are E. coli, transmissible gastroenteritis (TGE), clostridia, or rotavirus. Antimicrobial treatments, or antibiotics, offer some help, but the mortality rate for first litters remains high.

A large factor in determining the mortality of pigs is the level of stress on the sow. Stress can change the chemical composition of the milk and lower the quality of immunity passed from sow to newborn. A lack of immunity in the sow will also leave all pigs at high risk for infection and death.

Providing methods of rehydration, antibiotic regimens, and temperature stabilization will save some pigs. Looking for early signs of illness can make all the difference. Dry skin, sunken eyes, watery-looking stool, and lethargic demeanor indicate that the pig is in trouble.

Each successive litter will become increasingly immune to diarrheal causation, and the survival rate will improve drastically as the sow becomes more immune to the causes of diarrhea. Make every attempt to avoid reintroducing the infection by sterilizing the stalls between litters. Prevention is the best form of protection.

Breathing

Pigs struggling to breathe are truly involved in a life or death situation. As a pig owner, you must be proactive in ensuring you do all you can to resolve the issue. This book will cover the two more serious causes, but you should consult your veterinarian in any instance of a serious medical emergency.

Labor stress

Long, intense labor will stress a newborn. The extent is unclear until the birthing process is complete. If the pig is not breathing, open the mouth and clear out any mucous that might clog the airway using your fingers.

Remove any remaining sack membrane covering the baby pig, and gently pull them up by the back legs. Swing them slowly between your legs in a pendulum motion. This will drain any remaining mucous and fluid.

Clear the nose and throat once again to allow air passage. Closely monitor any pigs that have a hard time breathing initially, as they are susceptible to pneumonia. Keeping them warm and dry will assist in their full recovery.

Anemia

A main ingredient of red blood cells is a protein called hemoglobin. It is charged with the duty of carrying oxygen-rich blood to the cells and bringing carbon dioxide back to the lungs to be expelled. Iron is a critical part of the composition of hemoglobin. A dietary and environmental lack of iron causes serious and potentially fatal problems.

Anemia due to acute iron deficiency is capable of causing fatal breathing difficulties in young pigs. Baby pigs grow at an astounding rate. The faster a piglet with anemia grows, the more at-risk they become. Labored breathing or diaphragm spasms called **thumps** are indicative of serious anemia in pigs.

Pigs kept in concrete confinement have no access to natural iron found in the soil, which leaves them iron deficient. Rooting is nature's way of allowing the sow to stock up on needed iron before giving birth. Without the additional iron in the sow's colostrum and milk, very little will be passed on to the newborns.

A natural fix would seem to be to simply add iron to the diet of the sow. Unfortunately, this does not work. Iron deficiencies have to be resolved within the individual pigs. This is not a hard task with small farms, but it could be a real problem in larger sow operations. It will make your choice of housing and containment that much more important. *See Chapter 5 for more information about this topic.*

Orally administering iron can be effective, but this is time consuming and hit or miss in end result. Common methods include applying an iron paste to the sow's udders, weekly doses of iron pills, or soil added to the creep floor, which is where the young pigs spend their time when they are not nursing so they will not be crushed by their mother — although soil should be completely free of any swine parasites and eggs.

Iron injections offer more than enough protection against anemia if given in the proper form and manner. Iron-dextran or iron-dextrin complex is the best form on the market to date to treat iron deficiency anemia. The recommended dosage is one injection of 100 milligrams for pigs weaning at 3 weeks and 150 milligrams for those being weaned after 3 weeks. You should give the shots intramuscularly — in the muscle — in the ham — rear — or neck muscle. Clean the injection site well with an alcohol swab before giving the shot to avoid infection.

Health-related Procedures

There are several procedures that can be done while pigs are small. These procedures should be done when the pigs are young because the younger the animal is, the less pain the pig feels during the procedure. This section will not attempt to answer the argument of humane versus inhumane but will explain the various common procedures performed on pigs.

SIDEBAR:

It cannot be stressed enough that processes such as these should only be performed by a qualified person. Use sterile equipment that is in good working order. Always remove the pigs from the immediate vicinity of the sow to avoid stressing her. Instinct will make her protective of her offspring.

Most procedures center around protecting the pigs from infection, although some involve sow comfort, piglet identification, and retaining the value of the animal. Not every farmer will need to perform every procedure, but certainly some combination of the listed procedures will be necessary. Weigh the benefits with the risks in order to make an informed decision.

There is no way to completely avoid stressing the pig when completing any of the procedures, but initiating them within the first 12 to 24 hours of life works best. Doing as many as possible in one session will help reduce the overall stress. Whether it is due to an immature nervous system or other factors, younger piglets seem less reactive to pain. This does not mean they do not feel pain; it simply means they react less to it and recuperate faster.

Many surgical procedures performed on older pigs, no matter how slight, are met with great resistance. The larger the pig, the harder they are to control. It can turn a one- or two-person activity into a multiperson challenge. It is much easier to perform these common procedures on young pigs when they are small, easy to control, and quickly recover.

Incorporate plans for the various health management measures needed before purchasing your first young pig. Consult with professionals who are willing to discuss the benefits to your animals and farm regarding each and every aspect of the following procedures. Talk with other pig farmers in your local area if you still feel unsure of what your pigs might need.

Castration

Castration takes away the ability to produce sperm so male pigs raised for breeding are naturally bypassed for this procedure. Barrows, or males raised for meat and show purposes, must be castrated in order to remove **boar taint**, or the foul taste of testosterone from the meat. Researchers are currently looking into ways to remove boar taint that do not involve physical castration because uncastrated males have better feed and growth efficiency than castrated males. But, castration does serve to reduce aggression in the herd. Castrating young male pigs makes it much easier to raise them in a group until they are old enough to go to market.

Castration should always be done while the male pig is very young to ensure minimal hormone output into the system. Too much hormone released into the pig's body affects both the taste and smell of the meat. It provides a pungent odor that is not well tolerated on the market. The prices for non-castrated males versus castrated are noticeably less at sale date. Older boars are sent to market later in life, but they bring much lower prices per pound.

Even though it is considered a straightforward procedure, castration should never be attempted for the first time without the presence and assistance of a veterinarian or qualified professional. Incorrect castration will cause damage and unnecessary pain to the pig. Watching the process a few times beforehand is helpful.

1. Lay the young male pig down on its side on a small hay bale.
2. Have someone secure the back legs to keep them from moving.
3. Make a small incision over each testicle, cutting through the scrotum. Use a sterile scalpel or castration knife for this.
4. Pull each testicle free of the incision along with some of the cord. The thin cord will be visible inside the incision.
5. Cut the testicle loose, or pull until the cord breaks.
6. Spray on a generous layer of antiseptic solution. No bandage is necessary.

Keep a close eye on the site of the castration for several days. Check for swelling, fever, discharge, or any other sign that it may not be healing properly. Keep pigs in a clean, dry stall to promote healing.

Developing hernias after castration is not uncommon, but a veterinarian should be consulted if this occurs. The problem is much too serious to self-treat. Tissue protruding through the incision or bulges is a sign that the area has herniated. The tissue will have to be pushed back into place and the incisions stitched shut.

Tail docking

Docking the tail on pigs most often ends up being a procedure based on personal farmer preference. Animal rights groups tend to vilify the process as being inhumane and unnecessary and the cause of undue stress and pain. The tail does contain bone, but in young pigs it is a very soft, gristle-like, or cartilage-fibrous consistency.

Bleeding when docking pig tails is minimal and often absent altogether. The market tends to be fickle regarding the value and desire of tail docking. Feeder pigs without docked tails may not sell at market.

Pigs that have to be fed in close confines are subject to having their tails bitten. This leads to infection, or worst-case scenarios of cannibalism and death. The pig with the full tail is viewed as a liability in this case, and many farmers will pass them up without hesitation.

There is a whole other segment of farmer that values the aesthetic over practical. Smaller farms have fewer problems with overcrowding in feeding areas and the preference is to have the tail intact. Checking into the whims of your local market will help guide you in making the best decision for your business.

Ear notching

Notches cut into the pig's ears are used as a means of identifying the litter and the number of said litter. When done during the first few hours after birth, the procedure seems to go virtually unnoticed by the pigs. A sharp, sterile cutting instrument should be used. Follow up using an antiseptic spray to prevent infection.

Check with your state on requirements for ear notching. Some states require this procedure for disease control and monitoring. Most pig breed registries also require ear notch identification before issuing pedigree certifications.

Navel care

The navel should be treated with an iodine solution after the birth of each piglet. Bear in mind that the navel will be in contact with the flooring and other pigs on a frequent basis. It presents an open door to infection.

Check often for inflammation or fever around the navel. The area should heal in approximately one week. Contact a veterinarian if problems arise. Infections can overtake young pigs quickly and cause death.

Wolf teeth

It will be important to don your dentistry cap and clip the wolf, or needle, teeth on each pig. Pigs are born with their teeth in place; these teeth are as sharp as needles and cause pain to the sow when the newborns nurse. It may become so uncomfortable for her that she shoves the pigs away. These teeth can also cause damage to littermates in the fight to suckle.

Pigs have two on the left front and two on the right front, both top and bottom. This makes a total of eight teeth. Use a sharp pair of clippers to trim these down to the gum line. Have an extra person hold the pig and help keep the mouth open. You will need a good visual on these teeth as they are clipped. Be sure to avoid nicking the gums because the pigs will have a hard time eating with a sore mouth.

Weaning

Weaning times and techniques have changed drastically over the years. The push for increased production has led to larger and larger pig-raising operations. From a commercial viewpoint, it makes more sense to raise more pigs, to wean them earlier, and to breed sows back sooner. However, this rapid production model does place added stress on pigs and can result in health problems and disease. Pigs are also more likely to react to stress and show aggression under these conditions.

Smaller farms and individuals have more flexibility in weaning. At times, the sows take care of the weaning themselves. This method guarantees a pig that is more well-rounded behaviorally, but they are not at any particular advantage health wise.

The larger you plan to grow your pig operation, the more expensive the operating costs will be. Financially, keeping your pig operation viable and on firm financial footing forces earlier wean times. Rebreeding becomes the critical pivot point to staying in business. Having sows produce two or three litters per year can make a big difference in your bottom line.

The laid-back approach to small pig farming is certainly less stressful, but profit is minimized by reducing the number of litters produced annually. Setting reasonable goals for you and your sows will reduce stress on the herd. Knowing where you want to be financially with your pig business will help determine a reasonable weaning time for all of your pigs.

When to wean

You can successfully wean pigs on a small farm by 8 weeks of age. You will start noticing the pigs beginning to munch on some of the sow's food by roughly 6 weeks of age. Once they begin to develop a taste for solid food, you can begin setting aside an area for weaning. To reduce stress on both pigs and the sow, pick an area where neither can see, hear, or smell one another. Lack of adequate planning will create unhappy pigs and a sow that refuses to breed again.

The initial weaning holding area should be relatively small. Something in the neighborhood of 4 feet by 6 feet is adequate space for an average size litter of ten pigs. A smaller area will be more comfortable for the pigs when first being pulled away from the sow. They will pile together seeking comfort.

Ensure the holding area is secure. The pigs will spend much of their free time looking for a way to get out and find the sow. Remember that fencing materials have to be designed to contain animals as small as shoats. Standard pig containment materials will have gaps wide enough for them to crawl through.

Use shallow containers of food and water when first introducing them to their new diet. They will need to be filled frequently, but pigs will not have any problems accessing the food. Add flavored gelatin to the water to encourage them to drink regularly.

Monitor the shoats closely to see that they all eat and drink enough. Expect a slight drop, or at least a stabilization of their weight, during the transition to the new diet. Tie any feeders open until they learn where the food is located and how the feeder operates.

Weaning a lot of pigs from numerous sows is a little more difficult because it takes more space, manpower, and determination. Large pig operations normally wean at 4 weeks of age. This is considered very young by any standard. The reasoning behind early weaning is to maximize the number of litters per year and minimize the exposure time of the pigs to germs. Larger numbers of pigs housed together will often end up experiencing disease breakouts of some type, at some time, during a given year. Minimizing exposure is a proven way to curb the financial disaster of diseases involving the entire herd.

These tiny pigs must be monitored closely for weight gain. Limiting their exposure to other pigs assists their natural immunities passed to them from the sow's colostrum. The environment has to be tightly controlled for temperature because they are still heat and cold sensitive at 4 weeks. Pigs weaned later than 4 weeks will normally be past the age of being so sensitive to changes in heat and cold, though they are still susceptible to outbreaks of disease.

Health and safety issues

Weaning healthy pigs safely should be the priority of every pig farmer. Monitoring weight, observing behavior changes, and noticing food or water intake changes may be the only clues you have of existing problems.

Keeping food and water available will not be enough to transition pigs to a new way of eating. There are numerous ways to add flavoring to both food and water. Try anything that will safely work. Be persistent because malnutrition, even over short periods, affects the overall health of the pig.

Socialize daily with the piglets. They can develop a fear of humans if you do not take time to interact with them every day. This does not mean you have to spend hours teaching them tricks, but they should be comfortable with your coming and going from their enclosure. Pigs can easily reach an excess of 300 pounds. It is easy to get hurt by a frightened animal that is in this weight range. Invest in your personal safety and spend a few minutes, two or three times a day, interacting with them. This may be difficult if you are planning a large operation, but it is just as important on a large farm as on a small one. Pigs struggle with behavior issues even more when weaned at a young age. The isolation is good for deterring disease but not so good for developing social interaction skills.

Steps that ensure that meat production is free from disease are important, but containment guidelines should always balance a concern for the environmental comfort of the animals. Federal regulatory committees and agencies frequently revisit these issues in a continuing effort to blend the necessary elements to produce a healthy food source. Pig farmers, no matter how large or small, should strive to meet and exceed these guidelines.

Picking a Young Pig

Once the decision is made to purchase a piglet, it is time to do some homework. Will the pig be raised for show? Will it be a family pet? Are you planning to raise pigs for food? The answers to all of these questions will help narrow down the search for breeds and breeders.

Potbellied pet variety breeders are easily found in most newspapers. You should always ask for references when dealing with a breeder who you have no personal knowledge. A few phone calls can save you a wasted trip if it turns out to be a poor breeder. Local Future Farmers of America (FFA) and 4-H Clubs can assist you in locating quality show pig breeders. You do not have to pay a fortune for a good show pig. There have been $50 pigs that have won major awards at state fairs. Veterinarians and feed stores offer a wealth of information regarding local feeder pig breeders. You want to find one that uses vet care when necessary. Make appointments to begin visiting numerous breeders.

Not every type of pig breed will be available where you reside. It may require travel if you are set on a particular breed, but long-distance travel stresses pigs. It might be better to settle on another breed that is raised close to home.

Purchase more than one pig at a time when you can. Pigs are extremely social and will do better with company. Take your time when selecting your pig or pigs. Each one will have a distinct personality. Good breeders will not rush you to make a decision. Even if you do not make the purchase, the pigs will appreciate the extra attention.

What to look for in a breeding farm

The next step will be making appointments to visit the farm and see the pigs. Do not purchase pigs if you cannot see the conditions in which they were raised. Purchasing live animals carries a certain amount of risk, but reducing these risks will give you a better chance for success.

Does the farm look clean? Is everything orderly, or is it in a state of chaos? Lack of routine and cleanliness are big warning signs that the animals are probably not top priority to this farmer. Pigs stress out when chaos reigns.

Ask a multitude of questions. Does the breeder seem receptive to your questions or evasive in answering? Find out about their weaning procedures and what type of food the piglets eat. Are they fed a medicated feed? Piglets on medicated food stress more when transitioned to regular feed. Any change in feed can cause some degree of stress for young pigs.

Here are some specific questions you should ask the breeder:

⑤ How old were the piglets, or how old will they be, when weaned?

⑤ Are the males castrated?

⑤ When were they last de-wormed, or have they been de-wormed?

⑤ Are there any older siblings of these piglets that you can see?

⑤ Can you see the sow?

Male piglets should be castrated unless you plan to breed them. Boars are somewhat harder to work with and are a little more aggressive in nature. Other than individual personality traits, there is not much difference between barrows and gilts. Keep in mind that it is best not to transport a pig on the same day it has been castrated because the pig has already been through some physical stress. Too much stress in one day can cause a setback for the pig.

Are the pens overcrowded? This contributes to higher illness and infection rates. Proper herd management allows for adequate space. Mismanagement in one area can signal unseen problems in another.

Is the breeder able to give specific genetic information about the piglets? If the breeder cannot identify lineage past the sow and boar, they have probably not owned the pigs long. It is important to know the genetics when you purchase pigs for breeding. This is critical if you plan to raise any type of pedigree pig.

Quality breeders will welcome any questions you have and normally offer more information than needed. If you have any doubts or qualms, keep looking. The perfect piglet is out there somewhere waiting for you.

CASE STUDY

Noni Mammatt
Australia

Noni Mammatt has raised pigs in Australia for 16 years. She began with a sow that farrowed a month later and presented her with 15 piglets. Through many twists and turns over the years, Mammatt has come to have a current operation of 130 sows. She has had as many as 400 sows and says that she will probably have around 500 sows again soon, along with a large number of grower pigs. Mammatt's main focus is selling mated or bred gilts.

Mammatt says that she raises pigs for several reasons. "I like the animal immensely and the habits and their curiosity aspect. If other people like pigs the way I do, then yes they should grow them. They are not like sheep and cattle. They come for a look instead of running away."

Mammatt suggests that others can get started raising pigs the same way she has done. "Do as I have done and get one that is mated and a few smaller gilts and a small boar, and give it a go. A little reading might be a good idea beforehand. Borrow books from the library and from older members of the pig industry. You also get some verbal experience from these people on how to feed and what to add when making up your own feeds and how to keep them warm and dry and lean for market."

Mammatt prefers to keep her pigs outdoors. "I run my pigs all outdoors... I have done this for 16 years. I have them in paddocks, and they are given grain. They also have access to pasture when the paddock is locked up for the season, and this is free feed so to speak. The pasture is only around knee high so there is plenty of protein in the feed, which will last a good amount; for 25 to 30 sows around five to six weeks of feed.

"I like the idea of outdoor production, as it is a healthy way to produce pigs. It is also a cheaper option than building an indoor unit."

Mammatt says she raises several breeds, along with one of her own creation. "I raise large whites, Hampshire, Berkshires, Landrace, Duroc, and red saddles, which are a little different from the norm. These are a breed of my own, which I have slowly developed, and am now getting pure litters coming through. They are quiet, produce plenty of milk, have good numbers, and don't seem to lose too much condition while they carry their young, and make great mothers, as well as fosters. Are good to handle also."

She says she feeds wheat, barley, and lupins to her sows via a selection of diets formulated by a nutritionist. "I then only have to follow what is given to me and add the grain that is also on the diet. I have experimented a little with feeds as well. I give hay for bedding, and the sows consume some of this as well. The growers also eat hay and this helps them to digest the grain feed."

As for tips on raising pigs, Mammatt said, "The old timers who used to grow pigs back when I was not around would feed their pigs beer if they saw a sow with problems during labor … this quieted the old girl down."

Mammatt offers these last words about how she raises her pigs: "Being small producers, we probably do things a little differently than our indoor counterparts. I grow with huts to keep them dry and shaded and ring lock and barb and a hot wire to keep my girls in their paddocks. I use sprinkler systems around the piggery for their wallows. I use nipple drinkers for them to drink out of, and for the growers, I use both a trough and a nipple drinker. I feed the girls on the ground, and for my growers, I feed them in self-feeders. I have a lane way to move pigs around in. I weigh my pigs in a shed to keep the water out of the scales and the hot sun from beating down on the pigs and me. I have a chappy who brings his truck to my place to load pigs and take to market."

How should a healthy piglet act?

Piglets are remarkably different in temperament and personality, but there are certain characteristics that are seen in all healthy pigs. Curiosity is an ever-present trait. They will scamper over to investigate when you enter the pen. Piglets are also incredibly energetic. Unless they are still unweaned, pigs should spend plenty of time playing with littermates, food, and whatever else they get a hold of. Piglets that isolate themselves or seem excessively sleepy should be avoided. These signs can indicate malnutrition or illness.

Healthy pigs should be wide eyed and alert, especially with an unfamiliar person in their presence. Disorientation or stumbling is often due to dehydration or neurological disorders. This is a serious problem in a newly weaned piglet. Mortality rates are high with long-term dehydration.

Pick the piglet up gently. Does it squeal loudly? This would be a normal reaction because pigs do not like being picked up for any reason. If one goes limp and quiet when you pick it up, there is most likely a respiratory problem present. It is best to pick a pig up by its rear legs. If you attempt to pick it up and cradle it like a baby, it will most likely squeal and squirm more.

Coughing, wheezing, and raspy breathing are red flags that the piglet has a respiratory infection. These are easily transmitted so caution should be shown when purchasing from this breeder. Untreated communicable disease in pig herds is hard to escape.

Take plenty of time to watch how the piglets play among each other. They reveal much about their personalities in the way they play and interact. Some will appear dominant, while others seem very submissive. Return at a later date if the piglets are not yet weaned. You will not be able to tell much about them while they still nurse. Normally suckling piglets are either sleeping or eating.

How should a healthy piglet look?

Healthy shoats are dynamic little animals. The one activity they enjoy most next to eating is playing. When a pig feels good, it spends a great deal of time in motion. The eyes should look clear and bright. Cloudy, puffy, or irritated eyes indicate cataract problems or conjunctivitis. The latter is treatable but very contagious. The snout should be clear of mucus and deformities. Check around the head and neck area for injuries or abnormalities. Inspect the lips and gums for discoloration or swelling.

Pick the piglet up, and inspect the navel. Redness and swelling indicate infection. Observe the area of castration on barrows. Abnormal lumps mean one of two things: a hernia or infection. Does the piglet get around well? Swelling in the joints and difficulty moving is not a good thing. Imagine the problems after the pig gains another 250 pounds. It is a good idea to inspect the hooves. Cracks or deformities make it difficult or painful to move easily.

Avoid purchasing piglets if they have marks indicating abusive care. Inordinate amounts of bruising and red marks could indicate that they are hit frequently. This would stress pigs out and cause an untold amount of physical and psychological damage. If you suspect the piglets are being abused, it is best to leave and report the situation to the proper authorities.

Summary

In this chapter, you have examined what you should know about piglets. You have learned the vital importance of colostrum and getting the newborn pigs off to a good start. You have also addressed some of the health problems that can occur in young pigs. You have also considered the health-related procedures that need to be done with your pigs when they are very young, such as castration for males, tail docking in some cases, ear notching, and wolf teeth removal. You have also covered the weaning process. Finally, you have considered what to look for in young pigs if you decide to purchase weaned pigs instead of breeding your own.

In the next chapter, you will consider your options if you would like to breed your own pigs. There is a great deal to know before you put a sow and boar together. You will need to consider your breeding objectives, for example. Do you intend to raise purebred pigs as breeding stock? Will you be raising pigs for market? Will you breed to keep some pigs for your own table and to sell some locally? Once you know your objectives, you will need to choose your breeding stock. You will consider choosing your breeders. You will also consider farrowing and sow care. There is still a lot to learn before any little piggies go to market.

CHAPTER 4:

The Birds and the Bees

You have already learned about raising piglets in the last chapter, along with choosing young shoats so you can start raising your pigs. It is often a good idea to begin your pig raising experience with one or two shoats and raise them for your own table or raise one and sell one in order to see how you like raising pigs. If you decide you want to continue raising pigs, there will come a time when you will probably want to start breeding your own pigs. If you are at the stage of thinking about breeding your own pigs, then this chapter is for you.

Everyone loves a happy ending, but placing a gilt and boar together is no guarantee they will produce piglets. Environment, health, stress, and diet play as big a role in successful breeding as having a viable couple. Whether you plan to breed right away or over time, it is important to understand how the reproductive system works in order to maximize the chance of healthy piglets.

Purchasing a pregnant sow is an alternative way of starting a herd if you want to avoid the time and headache of matching a gilt and boar. This process works, but it is impossible to estimate an outcome for the litter. The genetics are often sketchy at best, and the health and stress levels of the boar and sow during the mating process are unknown to the buyer.

Taking care of the litter begins long before a pregnancy starts. A healthy diet, stress-free atmosphere, and proper care of the sow before, during, and after pregnancy are the keys to continued healthy pig production. The better the effort you put into the overall breeding program, the better your results.

Choosing Your Breeding Objectives

Before you buy a gilt or sow for breeding or invest in a boar, it is important to consider what your breeding objectives are. People keep and raise pigs for many different reasons: to supply pork for their own table; to sell to market; to sell young feeder pigs for others to raise and slaughter; to raise purebred seed stock for other farmers to purchase; or to show some of your pigs in a 4-H Club show. Whatever your goals, it is best for you to have them clearly in mind before you decide to breed your pigs. This will help you make the best decisions each step of the way.

For example, if you plan to raise a few pigs for yourself and sell the remaining young pigs when they are weaned as feeder pigs, then it will be fine to crossbreed your gilt or sow to a good boar if you have one. Depending on the breeds and the cross, you can achieve a desirable crossbreed that produces good pork. You will likely find local buyers for your shoats if they are healthy and a good size.

If you plan to build up a large operation and want to sell to market, then you will need to carefully select the breed you buy with an eye on weight gain efficiency, litter size, how many piglets usually survive through weaning, hardiness, and other factors. You will need to choose pigs that make good economic sense and that offer the best return on investment.

Economics play a role in the decisions of all farmers, of course. If you plan to raise purebred seed stock, you will need to consult with your breed association first and find out about the market for the breed. Is there a strong demand for the breed? Are other small farmers buying them? Are commercial breeders using them as an outcross breed to bring in some diversity to their herd? Are farmers buying them to raise as feeder pigs? Are there niche markets for the pork from the breed? You need to investigate carefully before investing heavily in purebred herds because they can be expensive, especially if they are rare or heirloom breeds.

No matter what your breeding goals, you can find other farmers who are doing similar things. It often helps to talk to other farmers about your plans. If there are no swine farmers near you with similar breeding plans, talk to your county extension services agent. He or she may be able to provide some helpful advice. The Cooperative Extension Service of the USDA is an informal education program the U.S. government offers through state and county governments and the system of land-grant universities. They offer information about agriculture, 4-H Club pursuits, food, the environment, and other matters of community interest. The program exists throughout the United States. You can find your local county extension service by checking under state or county government in your local phone book or by checking online for information about the program and your county. You can also find your nearest Extension Service office by visiting the USDA website at **www.csrees.usda.gov/Extension/**.

CASE STUDY

Walter Jeffries
Sugar Mountain Farm, LLC
Orange, Vermont
http://SugarMtnFarm.com

At Sugar Mountain Farm, Walter Jeffries and his wife, Holly, have two breeding herds of pigs with around 40 sows and four boars. Their pigs are a cross of several heritage breeds. They are mostly Yorkshire with some Berkshire, large black, Gloucestor Old Spots, Tamworth, and Hampshire, for good measure. Jeffries calls them Good Old American Pigs (GOAP). They have been selectively breeding the pigs for 12 pig generations, or seven years. As Jeffries explains on his website, "Gradually over time, this results in the improvement of the herd, stronger animals adapted to our climate, better meat flavor, marbling, length, temperament, mother, and pasture grazing ability to name a few of the traits we select for."

Sugar Mountain Farm raises the pigs from farrow to finish. They deliver their fresh, all-natural pastured pork weekly to local stores, restaurants, and individuals year-round. Jeffries gives several reasons for raising pigs: "We're good at it. They fit well with our farm. We enjoy pigs. We love pork and pork sells well. We used to raise sheep but they don't pay the rent, so to speak. We also tried meat chickens and rabbits. Pigs bring home the bacon and pay the mortgage."

Jeffries says that pigs are an easy animal to raise, especially during the summer. They can pasture, eat a wide variety of foods, and produce excellent meat and lard. "Raising them over the winter as we do for year-round markets is a lot harder but that isn't necessary for the homestead pig. People can stick with the easy summer months. Pigs and chickens are both good starting animals, and they mix well, complementing each other on the farmstead."

For anyone interested in getting started with pigs, Jeffries suggests planning ahead. "Get two to four weaner pigs in the spring, and raise them over the summer rotating on pasture. It is critical to reserve piglets early, as they sell out in the spring. We already have deposits from people for next year's piglets."

For the pigs at Sugar Mountain Farm, Jeffries likes to raise them on pasture as much as possible. "We raise our pigs free-range, outdoors on pasture using managed rotational grazing methods — same as with sheep. We currently pasture them on about 70 acres. The vast majority of our pigs' diet is pasture/hay. They also get dairy (primarily whey), apples, pumpkins, etc."

When it comes to slaughtering, Jeffries knows a great deal. "We sell to stores, restaurants, and individuals, which requires inspected slaughter so each week, we take ours to a USDA-inspected facility. We are in the process of building our own on-farm state/USDA inspected slaughterhouse and butcher shop. We have trained for 18 months at commercial meat cutting in preparation for this." You can check his website for more information about the on-site butcher shop they are building.

Choosing Your Breeders

Once you have decided what your breeding goals for the future are, you will be better able to choose your pigs for breeding. If you are just starting out and you are not yet familiar with the other pig farmers in your area, talk to your veterinarian, your feed store owner, and your county extension services agent about good local pig breeders. If you plan to purchase purebred pigs, check with your breed association for help in locating the nearest breeders. Get in touch with some of the breeders that are recommended to you to set up a time to visit. You can also visit breeder auctions and sale barns to find gilts and boars for sale.

When you start to look for your breeding stock, there are some important things to keep in mind:

⑤ Start planning well in advance. If you have gilts or sows ready to breed, do not wait until the last minute to look for a boar for them. You should not expect your fellow farmers to drop everything so you can get your pigs bred at the last minute. You will also have a better selection of breeding partners if you begin searching early.

⑤ Study pig genetics and pedigrees. Pedigrees are particularly important if you will raise purebred hogs. By studying pedigrees, you will come to understand what each particular pig contributes to a litter, and you can make the best possible breeding decisions.

⑤ No pig is perfect, not even yours. There is always room for improvement in each generation. Try to choose a boar for your gilt or sow with the goal of improving some things about her. However, you should keep in mind that genetics are complex, and you will not be able to drastically change many things in one generation. Instead, focus on trying to improve one or two traits in your gilt or sow with the boar. With a little genetic good fortune, you should have a litter of piglets that will be better than their parents.

⑤ When you visit another farmer's farm to look at gilts or sows for sale or to look over a boar for breeding, ask if you can see as many relatives as possible. It is always a good idea to see parents, offspring, siblings, and other closely related animals. You are not just buying a sow or gilt, you are buying their genetic contribution to your own herd. That means that the genes found in their relatives may also show up in your future pigs so it is best to get a good look at them to make sure you like what you see.

⑤ When you visit another farm with the purpose of looking at breeding stock, make sure you ask plenty of questions. Ask how the pigs are being raised, what they eat, if they have had any diseases, how they have been wormed, if they were bred naturally or by artificial insemination (AI), and anything else you think is pertinent. It is much easier to ask these questions before purchase than to have to bring a pig back because of something you failed to ask.

⑤ You will want to keep your eyes open and observe the farm. Is it well kept? Do the pigs seem happy? Mud and manure are normal parts of hog raising, but you will want to make sure that any pig you buy for breeding comes from a farm that is not prone to disease. You will also want to make sure the pigs you buy have not been unduly stressed because this could affect their fertility.

⑤ Finally, you should try to obtain breeding stock that comes from a farm similar to yours. It is easier for pigs to adapt to a new place if they will be going to a farm that is operated using the same methods. For example, if you buy a sow that has been raised in a confinement system, she will have trouble adapting if you pasture raise your pigs. Choose sows that are already used to the kind of methods you use. That way they will not be stressed, lose weight, or be slow about coming in season.

Your eyes should be able to tell you if the gilts or sows are in good flesh. As with young pigs, gilts and sows for breeding should have bright eyes and be lively, with no signs of illness. You should select the largest, most well-developed gilts available. If you are breeding purebred pigs you will need to know more about the correct conformation for your breed, but in general, a good gilt will have a good body capacity — in other words, she will be long and wide in the middle. This will give her plenty of room to carry a large litter of piglets. She should have at least six pairs of evenly spaced mammary glands. Some breeders are now selecting for more — 14 and 16 teats are not unusual — but the important thing is that the gilt or sow should have plenty of room for the piglets to nurse. She should also have a somewhat level topline or spine along the top of her back, which will contribute to staying physically sound. Her hooves should be big and well formed. Ideally, your gilt or sow will have large hams as well.

The trend today is toward longer, leaner pigs instead of the cobby or rounded pigs often seen in the past. Pork today is about 30 percent leaner than it was in the 1950s because of selection for leaner pigs and different body types. It compares favorably with chicken and beef in terms of calories and fat.

The Female Reproductive System

The female pig reproductive system consists of a right and left ovary, one cervix, uterus, oviduct, vagina, and vulva. The internal sexual reproductive organs are located beneath the rectum in the abdomen. They are supported internally by a strong connective tissue that thickens during pregnancy.

The ovaries are at the center of pig reproduction. They provide the hormones and eggs, making reproduction possible. The pituitary gland in the brain stimulates the follicle stimulating hormone (FSH) and luteinizing hormone (LH) into action. These hormones are responsible for stimulating reproduction in the female pig.

The **follicles**, the hair-like structures that cover the surface of the ovaries, are stimulated to produce an egg with these hormones. Though there are thousands of follicles, a normal female will only produce ten to 20 eggs at one time. The eggs are released at **estrus**, which is the time that the female is receptive to the boar. This can be tested by pushing down firmly on the backside of the female. If she "stands" in a breeding position, then she is in true estrus.

Problems with ovulation occur most often when gilts are too young, sows suffer extreme weight loss during lactation, or there is the presence of cystic ovarian disease. The follicles will not fully develop eggs or release them. These are all problems that need to be addressed because they will directly affect litter sizes. Once the egg ruptures free from the ovary, the follicles change to producing luteum cells. These cells secrete progesterone and bring estrus under control. The uterus then becomes receptive to the pregnancy.

Fertilization of the eggs happens in the tube that connects the uterus and ovaries. This tube is called the **oviduct**. It carries the sperm in one direction and the eggs the other way. They meet at the halfway point, called the **ampulla**. It is better to have sperm waiting for the eggs because the egg can only be fertilized within 12 hours of its release. The sperm needs to be introduced by natural or artificial means to the female once you determine she is in estrus.

Before breeding, you should make sure your gilt or sow is current on vaccinations that are given in your area. There is a wide range of vaccinations that can be given to pigs, and not all of them are given in all parts of the United States or Canada. It is best to consult with your veterinarian or county extension services agent to find out which vaccinations are recommended where you live. There is no need to stress a gilt or sow by giving her unnecessary vaccinations. At the same time, it is important to provide her as much immunity as possible to the diseases in your area so she can pass that immunity along to her newborns.

The Male Reproductive System

The boar's reproductive system is comprised of the penis, the two testicles, the scrotum that covers the testicles, the urethra, and the vas deferens. Although the penis and the scrotum are the most visible parts of the male reproductive system, the vas deferens and the urethra, which are unseen, are important in moving sperm from the testes to the penis and delivering the sperm during reproduction.

The **vas deferens** is a muscular tube that lies above and behind the testes, close to the urethra. During mating, the vas deferens helps propel the sperm in the testicles up into the urethra. From the urethra, the sperm can be ejaculated out of the penis and into the sow's vagina. The testicles rest inside the scrotum or scrotal sac. The sac is a thin, muscular skin that is flexible. It has fibro-elastic tissue inside. The sac serves to keep the testicles at a steady temperature to aid in sperm production and to provide protection for the sensitive testicles.

The boar's penis may be visible at times, or it may be sheathed in the prepuce under the boar's stomach. The **prepuce** is a pouch of skin that hangs under the boar's stomach. During mating, the boar's penis becomes rigid because of an increase in blood pressure. Male sperm production is a continuing process. As groups of sperm cells leave the testicles, new sperm cells are produced to take their place. They are stored in the scrotum to mature for up to seven weeks. It is a constant replacement process that ensures that a healthy male always has available sperm. Stress from a difficult living environment, poor diet, excessive heat, and illness will greatly reduce the sperm count in a male pig. Rectifying these situations in a short time will help the male recover sperm count quickly. Long-term stress permanently reduces sperm counts.

The benefit of the male reproductive system is it is visible to the eye. Problems, such as abnormal swellings, damaged penis, and size variance, in testes are very obvious. This makes shopping for a boar that much easier.

Building Your Herd

Building a herd is easily achieved through a variety of methods. You do not necessarily have to purchase a gilt and a boar and start a herd from scratch. Thoroughly researching the local market should turn up pig farmers who will sell pregnant sows. The drawback to this method is in not knowing if the genetics are as stated. There is also a question of diet and stress levels on the pregnant sow. A veterinary check can suggest whether the sow is healthy overall, but it still can leave questions as to the long-term care she has received both before and after being impregnated. Ask for references when embarking on this type of herd building.

It is best to buy more than one sow from the same farmer when possible. Buying several sows from different farms will subject the sows and piglets to germs they are not used to. Pregnancy and birth are not great times to take the chance of introducing illness to your herd. Do not transition the pregnant sow to a new type of food. Make sure you have the same type of food on hand that she currently eats. Dietary stress will show in weight loss, poor milk production, and the possibility of miscarriage.

Avoid purchasing gilts that are first-time moms. The stress of a first pregnancy compounded by a change in residence might cause the female to become aggressive to the piglets when they are born. The survival rate for these piglets is slim without intervention. This adds to the variable mortality rate that already exists for first-time litters.

Traditional breeding is sometimes dangerous for the gilt. Overly aggressive boars can kill a young female. Mixing a little traditional breeding along with purchasing already pregnant sows will give you a chance to develop good pairs without compromising initial herd growth.

Finding a sire service versus keeping a boar

Keeping a boar can also have its own challenges. Some farmers prefer to keep a herd of sows and gilts and use a sire service when it is time to breed. A **sire service** is essentially a stud service available for a fee. It can be easier to manage a single sex on a farm instead of dealing with both adult males and females. Boars can be somewhat aggressive and difficult to handle. And, using a sire service allows you to choose boars with a different genetic contribution for each mating instead of repeating your breedings, which can be good from a health and genetics viewpoint if you intend to keep some of the pigs in the litters for breeding later.

If you choose this approach to breeding, you will need to do some homework before you plan to breed to ensure you choose a healthy boar that complements your gilt or sow physically. If you raise pedigree pigs, then you will need to look for a boar of the same breed. If you have a popular breed in your area, you will probably have many boars from which to choose. However, if you have one of the rarer breeds, it is best to contact your breed association for help in locating farmers who offer a sire service for your breed.

If you intend to use a boar on a nearby farm, you and the other farmer can work out a good time for breeding. You may need to take several of your sows to the other farmer's farm for a few weeks or the farmer may bring the boar to your farm for a stay. If your preferred boar lives in another state, however, you may need to consider using artificial insemination to accomplish the breeding.

If you are interested in keeping boars so you can operate a sire service, you should begin with one or two proven young boars rather than starting with male weaner pigs. This will save you considerable time because you will already know they are nice specimens, capable of siring some nice piglets. You should choose young boars that have the pedigree and genetics you are looking for, whether they complement your own herd of sows or they match the kind of pigs raised in your area. Your boars may be in high demand if they belong to an uncommon breed in your area and they have good genetics. If they are boars from a heritage breed, you may have to advertise them so farmers from other parts of the country will know what kind of boars you have available.

Choose the best young boars you can afford in terms of genetics and physical attributes. You can find boars for breeding at fairs and good livestock shows and from other breeders who supply pigs for breeding purposes. Boars can be a considerable investment so it is a good idea to take an experienced hog person with you when you visit a farm to make decisions about purchasing them. If you plan to operate a sire service, the boars will be your largest purchase. Do not try to save money by purchasing second-rate animals. You will be known by your boars and the kind of piglets they sire so choose good animals even if they cost a little more. Prices for breeding boars vary depending on the breed and area of the country. If you are selling a "dose" of boar semen for artificial insemination of a sow — the term for a single collection of semen — prices range from $50 to about $300 and more, depending on the breed, its scarcity, and the desirability of the sire.

Of course, even if you have your own sire service and you keep boars for others to use at stud, you will probably keep your own herd of gilts and sows. This will be a good way for you to watch for promising your male piglets that may join your boars in the future as good sires. You can keep the best gilts for your herd, too, and sell off the other young pigs as breeding stock for others to raise. You can also keep some pigs to raise for your table or raise some pigs for market. If your boars produce well, you could have a good demand for breeding stock by taking this approach, especially if you have one of the heritage breeds or a breed that is in demand for hog shows. Breeders also provide many litters of well-bred pigs each year for 4-Hers and FFA members looking for young project pigs to show. You will have to be careful that you do not castrate all of your young male pigs too early if you plan to consider some of them as future boars. Castration would put an end to any breeding career.

Do not breed a young boar too often. You will deplete his sperm and reduce his sexual drive. As the boar gets older he can mate with more sows per week. In general, a boar that is 8 to 10 months old can mate once per day, or about five times per week. A mature boar, more than a year old, can mate twice a day on occasion, and may mate up to seven times per week. Collecting the boar's semen for artificial insemination would be considered the same as a mating.

If you do keep more than one boar, you will need to keep them penned separately. Boars normally do not get along with each other so you can expect aggression and serious fighting if you try to keep them together.

Artificial insemination versus introducing a boar

There are farmers who will preach heavily against artificial insemination (AI) over natural breeding, but there are positives to both methods. AI is known for being difficult with low yield results; yet, natural breeding takes time and can sometimes prove futile. However, artificial insemination is much more precise now than it once was. With good timing there are often successful matings and litter sizes are comparable to those of natural breedings. Trial and error tends to be the best method. Keeping an open mind about what it takes to produce successful litters is more important for growing your business than following trends.

When introducing a boar to a gilt, you should contain them near one another for a month or more if possible. Mother Nature will then take over, and the gilt or sow will be stimulated into estrus over time. Set aside four or five gilts per boar, if possible. It is easier to start a boar with the maximum number of gilts rather than try to introduce young females with sows that are already breeding with the boar. Jealousy issues will turn the situation chaotic quickly.

The best reason for AI is in situations where you have purchased a boar that is from a good bloodline and he turns out to be overly aggressive. The best protection for your females is to artificially inseminate them with the sperm from the boar. The success rate should be higher than normal if the boar is kept near the females. The intact male will bring on estrus, and it will be less of a guessing game as to when exactly to inseminate the female. In fact, if you work with your veterinarian, you should be able to pinpoint the time of your sow's ovulation very closely. Experienced farmers will be able to tell the best time to breed or to perform an AI by noting their sow or gilt's physical changes. If you plan to perform an artificial insemination, it is best to work with your veterinarian or with an experienced pig breeder in the beginning so they can show you what you need to know.

Artificial insemination is also very useful for farmers working with pedigree pigs, especially when the nearest boar may be several states away. It is not difficult to ship semen from a boar on one farm to a sow on another farm by following careful procedures. In these cases, AI makes it much easier to maintain heritage breeds and small herds of less popular breeds.

Using artificial insemination can also give you access to a diverse range of bloodlines from all over the country, or even the world, without the trouble and expense of keeping boars on your property. It does require proper timing and technique to achieve a successful breeding using AI, but your resulting piglets may be worth more because of their superior genes.

Whether you use a sire service from another breeder, artificial insemination, or one of your own boars, it is important to have a brucellosis test done on the boar you will use. Brucellosis can wipe out entire herds, and it is sexually transmitted. Brucellosis can cause inflammation and abortion in sows and swollen testicles in boars. It can cause infertility and sterility and can also cause paralysis in the rear legs. Testing for brucellosis is required by law in some circumstances, and it is best to take no chances with it. The cost of the test is very small and a veterinarian can easily perform the test. You should check with your county extension services agent to find out the state and local requirements for brucellosis testing in your area.

Feeding for Breeding

Accomplished pig breeders will be the first to proclaim the importance of feeding smart. Boars do well if maintained on a regular schedule of 6 pounds of daily food intake. The needs of sows before, during, and after pregnancy can be a little tricky. Once you understand the reasoning behind the dietary fluctuations, it makes a little more sense.

A good diet is a major factor in maintaining the health of your breeding hogs. Optimum breeding performance can only happen if the hogs are healthy and happy. Developing good feeding habits that your pigs can depend on will produce good results.

The healthy function of the pig's reproductive system depends on good genetics and good food. Though there is limited control over genetics, you are in the driver's seat when it comes to food. Read the ingredients of any food before feeding it to your herd. Feeds that contain nearly all fat-building ingredients are not the best to offer your herd. There needs to be a balance of protein building blocks to promote healthy gestation.

The more litters produced annually, the higher the toll on the health of the sow. Feeding for pig breeding involves much more than daily sustenance. It is just as important to consider replenishing nutrients that are drained in producing each and every litter of pigs.

Gestation

The average length of gestation for sows is 114 days. During this time, the sow should be fed 7 pounds of quality mixed food per day. Weight gain is necessary at this time to prepare for feeding the piglets. The proper percentage of protein and other nutrients for various stages of growth will be discussed in Chapter 6. Avoid commercial foods that add antimicrobials, as they are passed on to the pigs through the milk. The overuse of antimicrobials in food animals is believed to be a contributing factor in creating treatment-resistant strains of bacteria.

A better solution to illness prevention is to feed the sow a healthy mixture of mixed grains, alfalfa, red clover hay, and table scraps. The added minerals and vitamins will build the immune system and create a nutrient-rich milk for the piglets.

Roughly 50 percent of the diet should consist of mixed grain. Too much alfalfa or table scraps will cause loose stools. This can lead to dehydration issues and affect the milk production. Field grazing is encouraged as long as it is not close to her due date. Maintain the 7 pounds per day of food, along with the grazing. The energy expended will cancel out the calories taken in.

Dietary modifications

Modifications or manipulations of the diet are frequently made in an effort to boost the number of pigs per litter. Temporary reduction in food intake four to six weeks before mating can raise the piglet number by one or two per litter. This is a remarkable number over the entire breeding career of a sow.

The normal feeding routine calls for 6 pounds of feed per day. Reducing this amount to 4 pounds during the weeks immediately preceding pregnancy has this strange effect on litter size. It should never be done long term, and the food should be increased once the sow is pregnant. The food mixture should ideally contain added vitamins and minerals.

The food intake should be brought to 7 pounds per day once pregnancy is achieved with an added pound per piglet once she nurses. Dietary restrictions during gestation will cause harm to the litter and the sow. An increased diet after pregnancy will ensure plenty of rich milk for the newborns to drink. The diet can return to the normal 6 pounds per day once the piglets have been weaned, provided the sow is in good shape. Make sure she is fit before breeding again.

Keeping the sow happy

A happy sow is a healthy and well-adjusted sow. Pigs enjoy clean and comfortable surroundings. Good food and a clean water source are musts when attempting to breed. All water troughs should be filled with clean water daily. The food should be placed in troughs or easily accessible self-feeders. The less she feels she needs to "fight" for her food, the happier she will be. The benefits to keeping the sows happy are reflected in their eagerness to reproduce young. A female that feels she has to compete for food, water, sleeping areas, or attention grows stressed. The end results will be illness, decreased litter sizes, or high mortality rates in piglets.

The use of high-quality food is important. This does not always mean the most expensive. There are actually brands of very high quality that do not cost a fortune. Nor does it mean feeding your pigs swill or garbage. It is important for you to learn what is in the food you feed your pigs and to buy nutritious food for a fair price. It is worth the investment to grow a healthy herd. But, you certainly may feed your pigs healthy food scraps from your own kitchen or vegetables from your garden. Your pigs will enjoy them.

Adding bran to the feed in the week leading up to the birth of a litter will help boost the comfort level of the sow. Constipation is not only bothersome, but it can also cause fetal distress to the piglets at birth. It is best to avoid those problems altogether. A nutrient-rich diet during pregnancy and suckling keeps the sow healthy. Avoiding large drops in weight during this process reduces stress on her both emotionally and physically. She will be more likely to take her time when feeding piglets if she does not experience hunger pains herself.

Pigs are not fond of changes. The same can be said of dietary concerns. If you must change food brands, do it, and do it no more. Shopping around for the deal of the week will not sit well with a sow. Consistent feeding with the same food is more beneficial than the couple of dollars you can save by brand flipping.

Farrowing

No matter how many piglet deliveries you experience, there is always excitement in the air when a new litter is on the way. Farrowing, or giving birth, is a natural process that needs close supervision. This is especially true with gilts experiencing a first litter. Problems are not a common occurrence, but when they crop up, they can be serious.

You can do a lot to prepare for a safe birth event. Proper quarters and comfort will go a long way in keeping stress to a minimum. Assisting the sow with the piglets as they are born will help her focus energy on finishing the farrowing process. The female should be treated both inside and outside for parasites a week or two before the farrowing begins. Parasites can wreak havoc on the health of newborn piglets. Your veterinarian can provide internal parasite controls that are safe for pregnant pigs. If you are raising your pigs organically or trying to avoid chemicals, you will need to use something along the lines of garlic or rosemary as a wormer, though you should be aware that these herbs may not get rid of all of your sow's worms. If you plan to use a stronger natural wormer, then you should consult with a holistic practitioner to make sure you do not use anything that could harm the piglets. You will also need to control external parasites on and around your pigs by keeping their bedding changed, removing manure frequently, and treating for any pests bothering your pigs with either natural or chemical methods.

The human presence is vital in making sure the piglets are breathing and feeding and that the sow does not crush them because of confusion and pain. Mark the expected farrowing date on a calendar, and make plans to be around. It will be a wonderful experience you will not want to miss.

Preparing a farrowing site

In most states, commercial pig raisers use gestation crates for farrowing. The crates are steel structures that help prevent crush injuries in piglets during the farrowing process. The crates are very confining to the pregnant sows, though this method is used to keep the sows from lying on the newborn pigs and accidentally killing them.

Smaller farmers do not use gestation crates. Small pig raisers typically create a small pen for the farrowing female. If you plan to use a small pen, you should thoroughly clean and disinfect the area before use. Steam cleaning and spraying the pen down with a disinfectant spray will suffice. It needs to shine all the way down to the metal or wood.

Add a 3- to 4-inch layer of straw bedding to create a comfortable place for the sow to rest while giving birth. Set aside a corner area with a heat lamp to place the pigs as they are born. The goal is to try to maintain a draft-free ambient temperature of 80 degrees. This will be comfortable for both mother and babies.

You need to place the sows in the farrowing pen about a week before the due date. This will give their system time to gain immunity against any unfamiliar germs. She will then pass these immunities on to the piglets. Grab every advantage you can get.

Design the pen so you have plenty of light to see what goes on. *See Chapter 5 for more information on pens.* Emergencies will require quick action on your part. Easy access to the pen is essential.

Signs of labor

Pig breeders can make a close estimate as to when piglets are due, but at times, the sow can deliver early or late. Closely observing the sow over the week before the date should start yielding clues as to when labor has started. Here are some of the signs to look for:

- **Restless movement:** At the onset of labor, the sow becomes very restless. She cannot seem to find a comfortable position. She will frequently alternate between lying down, standing, and pacing the pen.

- **Swollen teats:** The mammary glands will swell as they fill with milk to feed the piglets.

- **Nest-building behaviors:** It will seem as if she is trying to prepare the area for the piglets. This involves moving the straw, scratching at the floor, and circling.

- **Vaginal discharge:** There will be a slight amount of vaginal discharge during hard labor. If it has a strong, putrid odor, you should contact your veterinarian.

- **Slightly elevated temperature:** Use of a rectal thermometer will confirm if the temperature has gone up. A slight fever is normal during delivery. The normal temperature for a pig is 101.5 to 102.5 F.

- **Circling and sniffing:** Increased amounts of circling in the pen and sniffing at her backside are strong indicators that birth is imminent.

Remove and wipe down the piglets as she delivers them. Place them near the teat once she has settled down to avoid injuring the piglet. Make sure she has passed the placenta, as this signals the end of the birthing process. Check the temperature after farrowing. If the temperature remains elevated to 104 degrees or higher, you will need to contact the veterinarian.

Signs of trouble in labor

The normal time for delivery of piglets is one every 20 minutes. Gilts take slightly longer because they are less experienced with giving birth. Prolonged labor tires the female. All of the following problems will require some type of intervention:

- Three or four hours of labor with no results when it has become obvious that the sow is experiencing a great deal of pain or pushing. This most likely indicates that there is a piglet in breech position or that a dead piglet blocks the birth canal.

- Foul-smelling or bloody vaginal discharge.

⑤ A partially visible baby that is blue in appearance and has not exited after some time.

If this is your first experience with a farrowing emergency, it is best to call the veterinarian for assistance. Self-intervention can be attempted, but you must make sure to give the sow a shot of antibiotic afterward. No matter how clean your hands and arms are, the chance of infection remains high.

⑤ Clean the vulva with antibacterial solution to avoid introducing germs to the birth canal.

⑤ Clip and clean your fingernails before assisting with the births.

⑤ Scrub your hands and arms with antimicrobial solution. Clean them up to the elbows.

⑤ Lubricate the hand and arm you will use. Dish soap will work if nothing else is available.

⑤ Bring all of the fingers and thumb together to a point and insert into the vulva. Enter slowly as the sow will be in fair amount of pain. Reach all the way to the area of obstruction.

⑤ Grab the piglet by the head, if possible. Pull it carefully toward you. Work with the muscle contractions to avoid hurting the sow more than necessary.

⑤ If the piglet is in breech position, firmly grab the back legs, and slowly pull the piglet out.

Oxytocin can be injected to stimulate the uterus in difficult labor situations. Your veterinarian can prescribe oxytocin for your sow prior to farrowing. It is not unusual for breeders to keep it on hand. Do not give more than 1 or

2 cubic centimeters. Overstimulating the uterus can be as bad as a difficult labor. Sometimes simply getting the sow up on her feet to change position or walk for a few minutes can shift a piglet into a better position and start labor moving again.

Post-farrowing Matters

The first 12 to 24 hours after farrowing are the most critical for the sow and for the newborns. Infection and stress-related problems will make themselves apparent within this time frame. Monitoring the progress of the piglets and the mother several times a day is important. The earlier problems are detected, the easier they are to fix.

The sow's diet will need to be increased slowly to avoid digestive problems. The sow's appetite will gradually pick up as she continues to nurse. The larger the litter, the more food she will need. One of the important goals of post-farrowing care is retaining the health and vigor of the female to make rebreeding possible.

Caring for the sow

The sow will be relieved once farrowing is over, but you will still need to monitor her for stress, unhappiness, and signs of illness or fatigue. With a healthy delivery, she should be up and about within an hour or two. She should be made to get up and move if she seems unwilling. Listlessness and disinterest in the feeding activities of the piglets are red flags that something is wrong. Monitor for any temperature spikes, which could indicate infection. The earlier she is started on antibiotics, the more positive the outcome. Check often that her teats are releasing milk.

The sow will need to be checked often for signs of mastitis-metritis-agalactia (MMA) syndrome, which is serious and needs to be dealt with immediately. **Mastitis** is hardened and painful teats that can be so bad that the sow refuses to feed the babies. **Metritis** is an infection of the uterus. Antibiotics will be needed to clear this up before she succumbs to toxins in the blood. **Agalactia** is a lack of milk. A shot of oxytocin will help stimulate the milk production. Do not attempt more than two injections of 1 or 2 cubic centimeters in a day.

If you have one sow that has delivered a good number of pigs and one that has only has a few, attempt to place a couple of the piglets with the smaller litter to decrease stress on the sow. This can be done by sprinkling each baby with talcum powder. The new sow will not reject them because all the babies smell the same.

If the sow initially shows aggression toward the piglets, pull them and keep in a warm area. Slowly reintroduce the piglets to her after she has calmed down. Her stress level and hysteria will decrease in an hour or two, and the piglets will still be able to receive the all-important colostrum feeding up to 24 hours after the birth. Consider fostering the pigs with another sow and hand-feeding colostrum if she seems unable to accept the litter. Most sows are very good mothers and instinctively know what to do when their piglets are born. Any confusion or rejection is usually short term.

Some pig raisers recommend keeping the farrowing site calm and quiet before farrowing in order to soothe the sow. You may wish to play soothing music during the farrowing. Talk calmly to the sow. It will also help keep the sow calm if the person present has a good relationship with the sow. All of these suggestions can help sows be more relaxed during farrowing, which, in turn, makes them more accepting of their piglets. These suggestions can be particularly helpful with gilts when they are first-time mothers.

Feeding a nursing sow

Nursing sows need to increase their food intake to compensate for the nutrients lost when feeding their young. You can provide the sow with a light meal soon after farrowing. Providing the sow with 3 to 4 pounds of feed plus a can of pumpkin to induce regularity is a good idea after farrowing. You can also include any of her personal favorites, such as vegetable scraps, to get her to eat. Then, gradually build up her diet again to her pre-farrowing meal intake — 7 pounds — plus 1 pound for each piglet she nurses. Adding bran to the diet post-farrowing is a common practice to avoid strain in the groin area from constipation. Constipation can be a serious problem at this stage and may make the sow stop eating or feel otherwise ill.

The generalized rule is 6 to 7 pounds of food daily for the sow plus 1 pound extra per nursing piglet. This needs to be cut to ½ pound per piglet if it is a litter with fewer than ten piglets. The food may have to be offered three times a day for her to have the time and appetite to eat it. It also depends on the appetite of the individual female. You will have to experiment some. If it is hot, the sow might eat better in the early mornings or late evenings.

Fresh water should be available to the nursing sow at all times of the day and night. Producing milk requires plenty of water. The water will also help prevent the sow from becoming constipated. The ultimate goal is to maintain the sow in good condition so she can rebreed quickly after weaning.

Breed Registry Agencies

When purchasing or raising a litter of purebred piglets, you will want to register them with the agency handling that particular breed within 90 days of their birth. The litter certificate will need to be filled out and sent in. Include copies of the registration for the parents.

Ear notching is the method used to identify the individual piglets. If they have not had the procedure done, you should request that the breeder have it done before you take the piglet home. Lack of ear notching will nullify any paperwork you receive for registration.

Pig registration has to be done by current members of the registry organization. If you are not a current member, you will need to acquire membership or ask the breeder to complete the registration for you. The fee for registering is usually nominal, but make sure the breeder is able to register the pig before purchase if it is an important issue to you. For example, if you raise Duroc, Hampshire, Landrace, or Yorkshire pigs, you can register them through the National Swine Registry. The cost of registering a litter that is 90 days old and younger is $12 if you are an NSR member. For litters that are more than 90 days old, the cost is $24 for members. Non-members can also register litters but the costs are double. If you own a boar of one of these breeds, you can have it on file with the registry. Sows and their litters can also be tracked with the sow productivity program. These programs allow the organization to follow things like the number of piglets produced per sow, the number of piglets weaned, and the breed's efficiency of production as it depends on number of pigs weaned per sow.

Why is registration important?

Registering purebred swine is the only definitive method that agencies and organizations have to keep track of the pig population. It shows the current trends in pig breeding and rearing by offering true numbers regarding growth of the herd, litter sizes, and breeding successes or failures. A real picture emerges of whether advances in genetic research and development work. The results can pinpoint problem areas that need to be worked on as an overall breed to ensure their survival.

The swine industry depends on knowledge of where pigs that enter the marketplace originate. Illness and disease can be traced easier by knowing the lineage of the hogs being sold. This helps protect both the market and the consumer. In addition, purebred pigs provide the genetic basis for commercial hybrids, which make up most of the pork production in North America, and for breed improvement. Purebred pigs also have distinct characteristics that could disappear without breeders and registries dedicated to promoting them.

Registering your purebred litter adds value to your herd. Pure bloodlines are worth much more than mixed breeds. It provides paper proof that the pigs are what you represent them to be.

Protection of rare and endangered species

There are breed registry agencies that work for the expansion of breeding rare and endangered species of swine. Many smaller farmers opt to set aside a small area to help keep some of these breeds from falling off the map. Tamworths, Berkshires, and large blacks, not to mention the Gloucester Old Spots and others, all have their fans, with good reason. These rare or heirloom breeds are maintained today mostly by small farmers who are intent on keeping the breeds going and showing the world their virtues.

Aside from assisting with the information needed to start breeding these types of pigs, breed registries maintain detailed databases regarding the successes of current breeders. It is encouraged to maintain a pure line of any rare breed of pig you may decide to raise. Mixing rare types of pigs with more commonly found varieties places them at greater risk of extinction. Every litter of mixed piglets is a missed opportunity for a pure bloodline.

Though it is a commendable endeavor, if you decide to breed a rare or endangered species of pig, do so with an understanding of the undertaking. It will take space away from your standard breeding stock. There may not always be a boar available for the sows. It may be tough to market extra piglets. If these are scenarios you can live with, then by all means, jump right in. You are the owner these special pigs have been looking for.

On the other hand, many small farmers do enjoy working with rare and endangered breeds of pigs. It is important to preserve these breeds for genetic reasons. Some of the breeds have historical importance and all of them provide genetic options to the current popular pig breeds that may be lost if these rare breeds are allowed to die out. In addition, farmers who maintain **seedstock herds** of the rare breeds, or purebred herds with breeding stock available for others to purchase, can find themselves in the position of supplying breeding stock to other farmers who wish to build up their own herds. The commercial pork producers also rely on seedstock breeders when they need to outcross their own herds.

Finally, heirloom breeds of swine have become very popular with many chefs and aficionados of the slow food movement who appreciate more flavorful food. The **slow food movement** is an international movement that has grown as an alternative to the fast food that people eat on the go. People who are interested in "slow food" are interested in many foods that are harder to find and that often take longer to prepare. They often prefer foods with unique flavors. Many of the rare, older pig breeds do not grow as rapidly as commercially raised pigs. They take longer to grow and their meat is much more flavorful than most modern pork. Pork from these breeds can sell very well at farmers markets and through direct marketing by farmers. Niche markets can be one of the keys to success with rare pig breeds.

The National Swine Registry

The American Yorkshire Club, the Hampshire Swine Registry, and the United Duroc Swine Registry joined forces in 1994 to form what is now known as the National Swine Registry. The American Landrace Association fell in with the NSR in 1998. As of this writing, the NSR represents the largest portion of the swine industry. Their genetic work is renowned worldwide.

A large concern for the NSR is in keeping the breeds genetically viable and pure. By consulting with owners regarding proper crossbreeding, they ensure strong, stable bloodlines. They help make and maintain a pool of pedigree swine by registering purebred litters.

Certified Pedigreed Swine

Formed in 1997, the Certified Pedigreed Swine organization is a combination of the Chester White, Poland China, and spotted swine organizations. Their goal is to maintain a registry of purebred swine to help preserve the purity of these three breeds. Genetic advancements in breeding methods are the consistent goals of the CPS.

Other registry organizations

There are numerous organizations that sponsor registration of various pig breeds. The process of registration is similar, although the fees may vary. *See the Appendix for a list of breed registry agencies.*

Summary

This chapter examined boars and sows and how to choose breeding stock. This chapter covered some of the goals people have when breeding and why they might choose purebred or crossbred animals. You have looked at farrowing and caring for the sow throughout the process. Finally, you have considered a few of the primary swine registries for purebred pigs.

For many pig farmers, breeding is at the heart of keeping pigs. However, in order to properly raise and care for your pigs you will need to consider housing, feed, and other aspects of pig care.

CHAPTER 5 :

Housing: Straw, Sticks, or Bricks?

Up to this point you have considered pigs: young shoats purchased from other farmers; piglets farrowed by sows; and sows, gilts, and boars purchased as breeding stock. All of these animals need a place to live when they come to your property. Whether you decide to build a house for your pigs, convert an existing structure, raise them in a pasture, or set up temporary housing, your pigs will need some kind of shelter.

While the three little pigs had to face a wolf huffing and puffing at the door, your pigs should have a much easier time of things. However, you will still need to decide what kind of structure to provide for your pigs. Many housing options exist for pigs today, depending on the size of your herd, your goals for raising pigs, and the kind of pigs you keep.

Pigs are extremely adaptable animals, which helps explain why there are an estimated 2 billion domestic pigs kept and raised at all times in the world. People have kept pigs in just about every kind of enclosure you can imagine, from nailing together old bits of wood lying around as fencing to using dog kennels for a sleeping place for their pigs and even to using a cleaned out oil drum lined with bedding as a sleeping spot. These may not be ideal situations, but, provided the pigs are well-fed and kept free of disease and stress, they should be happy and thrive.

No matter what your situation or how much room you have, it will probably be possible for you to keep a pig or two. Whether you already have a farm and you are looking to add a pig or two for your own table or this is your first farming venture, pigs do not take up a lot of space, and they provide good returns for your investment in feed and labor.

Structures

The most popular way to keep pigs on a small farm is to provide them with a covered, dry place to sleep next to an outdoor enclosure. This kind of setup allows the pigs to move around, even if it is a small space. They can seek out the shelter if it is hot or cold or if it is wet. They can be fed and watered in the outdoor enclosure. And, because pigs prefer to relieve themselves in one particular area of their enclosure, this setup will allow them room to move away from their feeding station and bedding area and keep the pen cleaner. This makes cleanup easier for you. It should only take you about five minutes a day to clean out the pen because you will do most of your mucking out in one small area.

If you choose to use this kind of setup, you can convert an existing structure, such as a shed, or you can build a new structure for your pigs. Or, you can opt for an entirely different kind of arrangement. For example, you could choose to keep your pigs indoors more, or you could choose to raise them on pasture.

Converting existing structures

Often the easiest way to begin keeping pigs is by converting an existing structure on your property, such as an unused farm outbuilding. A shed, a workshop, a lean-to, or an old barn can provide your pigs with the cover they need.

Ideally, you should allow a space that is 5 feet by 7 feet — 35 square feet — for sleeping space for each adult sow. Sows will need more room if they have litters: Plan on at least 8 feet by 8 feet — 64 square feet — for a sow and litter. For young pigs less than 50 pounds, you should allow about 3 square feet per pig. For pigs that you are growing out — from 50 to 200 pounds — you should allow about 8 square feet per pig. This assumes that the pigs will have access to a pen when they are not sleeping.

You can estimate 1 acre for two sows and their litters; 1 acre for 25 young pigs less than 50 pounds; or 1 acre for ten pigs that weigh between 50 to 200 pounds. If you have less than 1 acre of ground for your pigs, you can make adjustments in your herd to fit the room you have. The numbers provided here are enough for someone raising pigs for the first time.

When you convert an existing structure for your pigs, it is wise to have it enclosed on three sides, shielded against the prevailing winds in your area. One side of the shelter should remain open so the pigs can move into the pen for their feed and water. You may have to cope with whatever flooring already exists in the structure. Pigs do best on a hardwood floor. Wooden floors are easier on their hooves and legs than concrete and stay cleaner than a dirt floor. Concrete can also be a good choice, provided there are some grooves in the floor to keep the pigs from slipping.

Remember that floors tend to get wet and can become slippery so you will need to provide your pigs with plenty of good, clean bedding. Change out the bedding regularly to keep your pigs healthy and happy. Pigs traditionally enjoy straw as bedding, but you can also use wood chips or sawdust as a bottom layer. This is a good idea if you have a dirt floor because it will make the space a little easier to clean. Topping off the wood chips or sawdust with straw will provide your pigs with something to chew on and keep them very comfortable. Soybean stalks or dusty bedding of any kind is not recommended because they can cause respiratory problems. Aim for bedding that has good absorbency and that your pigs can rearrange as they like. That way, they can pile it deeper for warmth or spread it out to stay cooler.

Building a new structure

You can also build a new structure for your pigs. This can be a good idea if you are going into pigs "whole hog" and plan to raise many pigs or even if you simply plan to raise a modest number of pigs on a regular basis.

When building a new structure for your pigs, consider the location. You may be very fond of your pigs, but you will probably not want them to be too close to your house because of their odors. You should also consider factors like your weather and how much sun will shine on their location. Pigs do not like to be either too hot or too cold. Some nearby shade trees will be welcome during hot weather. You should also consider how much rain you get and whether your area is prone to flooding. Place the new building for your pigs on higher ground so the adjacent pen will not get soggy. If you have a garden, you may want to place the structure and pen nearby so you can discard some of your vegetables into the pig's pen. You should also consider access to your building because you may need to have pigs and feed delivered in large trucks and hogs hauled to the butcher.

Once you have chosen a good location for your new structure, you can begin building. Depending on how large the building will be, you may want to include some farrowing pens for your sows. It is recommended that the building be wired for electricity so you can have a heat lamp, heating pad, and other things that may be needed during farrowing or during a veterinarian's visit. A wooden structure is recommended if you are building from scratch. Metal buildings can hold in too much heat in the summer and cause your pigs discomfort. Good oak wood will last you for 15 years or longer. Make sure you lay a good foundation for the building that allows some air circulation beneath the structure because it is likely there will be some wetness that seeps through at times.

Walls in the building do not need to be made of wood because they can pick up dirt and be hard to clean. Pigs may also be inclined to chew on them. Instead, consider poured and reinforced concrete walls with an insulated interior. These walls will provide your pigs with more warmth in the winter and stay cooler in the summer. Ventilation is critically important in any building housing pigs so make sure your building allows for good airflow. You may need to provide some heat in the winter, depending on the severity of your winter weather. An electric heater or a propane heater can be used in the pig house if necessary, but you will need to take care that they do not cause a fire next to flammable bedding. For flooring, you can consider wooden or concrete floors. Wooden floors are believed to be warmer and easier on the hooves and legs, especially for large pigs. Concrete is easier to clean, and it will last a long time. If you do choose to go with wooden flooring, hardwood floors are recommended because they will last longer than other kinds of wood. Choose wood that is 2 inches thick to stand up to regular use from heavy swine.

The size and type of permanent structure you build will depend on how many pigs you plan to raise and what your focus will be. You can determine if you will need special areas for weanling pigs or if you need more room for finishing hogs. Building a new, permanent structure is an expensive proposition, but it may be worth the expense if you intend to devote a great deal of your business to raising hogs. You can obtain more specific building plans from your county extension services agent. The University of Tennessee has some excellent building plans online for many types of swine housing: **http://bioengr.ag.utk.edu/extension/extpubs/PlanList97.htm# Swine%20Plans**.

The mobile pig house

Because pigs are so adaptable, people have been very creative in coming up with interesting ways to house them. It seems that wherever there is some available land, someone has found a way to house a pig or two there, thanks largely to mobile pig housing.

Mobile pig houses cover a lot of different kinds of housing units. Quonset-type huts can be used successfully for raising pigs, as can hoop structures. Quonset huts are prefabricated, lightweight structures made of corrugated galvanized steel. They are all-purpose buildings that are easy to assemble, easy to move, and easy to clean. They come in different sizes. Hoop structures or houses have a steel framework and a clear, woven fabric cover. They are very popular with livestock producers who often use them to house livestock, to store farm equipment, and to store hay. Hoop shelters are also easy to assemble and maintain, and they are less expensive than permanent structures. Hoop shelters have the added advantage that they provide good ventilation and plenty of natural light. You can keep sows or weanling pigs in these units or even use them to finish pigs before butchering.

Farmers who pasture raise their pigs provide mobile pig houses that can be moved as the pigs rotate to different pastures or paddocks. For sows that farrow in the pasture, you can provide farrowing huts. Some of these huts are built of wood on an A-frame construction and slightly resemble a large doghouse. The sow can take possession of her farrow hut and deliver in a very natural way. These huts are easily moved and can be cleaned and used again as needed.

Pasturing

There have been times in our history and in many parts of the world when pasture raising pigs was very common. Today, most pork in the United States is produced by commercial pork producers who use a confinement system. There are obvious benefits to the confinement system: Pigs are maintained at the same temperature and in the same conditions year-round. They are not subject to the extremes of weather, which can affect their weight gain. This system also requires less human labor. In addition, pigs raised in these conditions are the most economical to feed in terms of weight gain efficiency. They grow fastest and reach their optimum weight sooner than other pigs.

Even small farmers practice a modified form of this system by using modified confinement to raise their pigs. Pigs are confined to a small pen or lot with a separate sleeping area. These pigs are exposed to weather fluctuations but they still enjoy many of the same benefits of confinement farming. They grow very quickly in these circumstances, and their weight gain efficiency is still very high.

At the other end of the spectrum is pasture raising pigs. With this approach, pigs are turned out in the pasture and left to grow. They grow more slowly than their confined counterparts, and they are exposed to the weather. Pigs cannot thrive on pasture grasses or forage alone so their diet must be supplemented with feed.

There are a number of pros and cons to pasture raising pigs.

Benefits of Pasture Raising Pigs

- Pigs grow more slowly — meat has more flavor, distinct from commercial pork.
- Less early investment in buildings occurs.
- Lower production costs for the farmer — labor, feed — exists.
- Permanent buildings are not needed.
- Pigs are under less stress — less cannibalism, less aggression.
- Manure is utilized in the pasture.
- The pigs' immune systems are stronger.

Problems with Pasture Raising Pigs

- Pigs grow more slowly — involves more costs to raise.
- Pigs may have an increase in internal parasites.
- Pigs can escape from fencing.

One thing that should be mentioned is that pigs that spend time in the pasture are normally very happy. Although that fact by itself cannot determine whether you raise your pigs in the pasture, it is an important consideration.

CASE STUDY

Peter Burrows, CEO
Brown Boar Farm
Julie Barber, operations manager
55 Lamb Hill Road
East Wells, Vermont
Peter.burrows@brownboarfarm.com,
julie.barber@brownboarfarm.com
www.brownboarfarm.com/
802-325-2461

Peter Burrows is CEO of Brown Boar Farm in Vermont. At Brown Boar Farm, they specialize in heritage breeds — Tamworths and Berkshires. As Burrows explains, "We raise pigs naturally without any nontherapeutic drugs, food byproducts, or any growth hormones or chemicals. We sell the meat commercially to restaurants, wholesalers, and directly to the public at the farm, farmers markets, or through other direct-to-consumer means. We also have our own commercial certified BBQ trailer that we can use for pig roasts at weddings or other catered events."

Brown Boar Farm has raised pigs since 2001. Depending on the time of year, they may have as many as 30 to 60 breeding sows, several working boars, and they may have more than 150 pigs running around when litters are produced. Burrows notes that heritage pigs do tend to have smaller litter sizes than other breeds and that not all sows are fertile; nor do all offspring necessarily live. As Burrows explains, if there is infant mortality, it usually occurs at birth or within the first two weeks of life. Most of the pigs that do survive after two weeks will grow to maturity.

Burrows says that he got started with pigs because of the terrain where he and his wife live in Vermont. A small part of their property contained some older pastures that they wanted to use to grow vegetables, but the pastures had become overgrown with briers and bushes. The rest of the 110-acre property is hilly and rock-strewn with deep forest. There is even a deep ridge on part of the property. An old-timer who lived nearby advised him that pigs were an efficient way to clear property. After researching pig breeds, Peter chose the Tamworth for his property. Tamworths have sturdy legs and love to be outdoors, and they are hardy enough for Vermont winters. They obtained six registered purebred Tamworth sows and one boar. After fencing off the areas they wanted cleared, they turned the pigs loose and discovered that the old-timer had been right. The land was cleared, the soil was better, and the sows were all mothers with healthy litters.

Burrows recommends that anyone interested in raising pigs do some research about the various breeds of pigs and the characteristics of each breed so they understand the pluses and minuses of each breed. He says that the heritage breeds are still pretty rare and purebred stock may be difficult to find. It will also cost more. "Due to the shortage of heritage breeds, someone interested in raising pigs may need to settle for what pig breeds are available nearby. Unless you have a heated barn, farms in the north cannot have births year-round so depending on where you live, you may not find pigs for sale every month of the year. If someone can find a pig in the spring to purchase, that pig could be ready for slaughter in the fall/winter and that timing would take the most advantage of naturally grown food."

Burrows also advised people to be realistic about how much space they have and how much room pigs need. "Anyone considering raising pigs should also decide how many pigs they want to raise and make sure they have prepared the space for them before they bring the animals home. Baby pigs are really cute and hard to resist, but they can grow quickly and anyone raising pigs needs to be realistic about the space needed and the fencing needed. Pigs need sturdy and reliable fencing to keep them from roaming and tearing up vegetable gardens and other plants you do not want them to. Pigs are also social animals, and they seem happier in groups, but the pigs do develop pecking orders, and they do not like to be separated from their groups. We find if we move pigs, it is easier on us and them if we do it as a group."

Pigs at Brown Boar Farm mostly live outdoors in pastures or semi-cleared forests where the sunlight can allow the growth of natural food on the forest floor. "Our pigs have shelter, but we find the heritage breeds prefer to be outdoors. Pig shelters do not have to be very sophisticated as long as the animals can find a dry spot and can have easy access. We have one large barn that we use for pregnant sows when the weather is getting colder. We can also use this as a group nursery as we can add heat lamps if we need to. We do not use farrowing cages and never restrict the movement of the mother. Confined pigs will sometimes develop anti-social behavior like tail biting and fighting, which often leads to pigs being even further confined."

Burrows says farmers can avoid such anti-social behavior by having plenty of separation and plenty of room outdoors. That requires a good fence design that makes it easy to move the animals from place to place. Burrows also points out that regulations on raising animals can vary from state to state. "Even if there are no regulations, you should be very considerate of your neighbors. You want to design your fencing to minimize waste and odors and be respectful of your neighbor's property. The strictest rules in raising animals for food are focused on handling meat. There can be different rules for the state and the federal level, and these rules cover both processing and labeling. If someone plans on selling meat out of state, then they must follow strict federal USDA requirements for processing and labeling. There are exceptions to these rules for on-farm personal use, but anyone considering selling and processing meat for sale to others must be very aware of the laws, as well as the need for food safety, and follow them faithfully. They should also make sure they have liability insurance that covers food handling."

Borrows is also active within the local farming community. "We also try to help out families that might like to raise their own pigs and slaughter them by hosting classes for people interested in learning how to do this. For example, next weekend, we are hosting a two-day workshop at the farm with the New England Organic Farming Association. We will donate a pig that will be slaughtered at the farm."

If you are interested in pasture raising your pigs, it is estimated that good pasture can provide between 30 and 50 percent of the nutritional value of grain. You will need to supplement your pigs' diets, especially if you have pregnant sows. *See Chapter 6 for more information about nutrition and supplements.* Farmers who become seriously interested in pasture raising pigs normally plant diverse fields of grasses and legumes so they can rotationally graze their pigs. This offers their pigs the widest range of nutrition and keeps the pastures from becoming overgrazed.

Fencing and Pens

Fencing and pens for your pigs do not have to be complicated, but they do need to be sturdy and well installed. Those cute piglets you bring home when they are 35 pounds are going to grow to around 250 pounds when they are ready for slaughtering. If your pens or fences have any weak spots, they will find them, which can mean an afternoon spent chasing pigs.

Pens are often part of pig housing with an indoor sleeping area for pigs. In other cases, pens can be used as a drylot. Just as it sounds, a **drylot** is a large, dry lot on dirt where pigs are kept. Whether you put up a pen for a drylot or as part of a housing setup, the pen needs to be designed for the pigs you intend to keep in it. If you use rails or planks, for instance, small shoats will quickly wiggle their way between them. If you use woven farm wire that is not strong enough, then some hogs will be able to push it over or push through it. Hog panels are a good choice to contain your pigs because they are designed with pigs in mind. **Hog panels** are all metal, woven wire and are easy to move. They also attach easily to metal posts, which makes them easy to put up. They are 34 to 54 inches tall by 16 feet long. You can buy them new or used. Used is fine, and you will save money if you buy them secondhand. You may also consider wooden slatted fencing if the slats are close together. This kind of fencing will keep shoats inside the pen and larger pigs cannot push through it. If you use wooden slatted fencing, you will need to make sure the slats are placed close to the ground so young shoats cannot crawl under the fence. You may wish to place some wire fencing at the base of the fence or run a strand of electric fencing around the bottom of the fence to keep small pigs from trying to dig their way out. Like many animals, pigs think the grass is greener on the other side of the fence.

You can purchase hog panels at the Tractor Supply Company® store (**www.tractorsupply.com/**) and at feed supply stores. You may be able to find hog panels for sale secondhand on Craigslist or in your local newspaper at a significant savings. You can also check with local farmers to ask if they know anyone who may have some to sell. You may also find hog panels for sale at livestock auctions. You will be able to find wooden slatted fencing at most home supply stores, such as Home Depot®, or you can go to a lumber yard and purchase the lumber to build the fence to your specifications.

You can also use electric fencing to keep your pigs in place. It is economical — a 4,000-foot coil of electric fence wire is $100 and a typical AC electrical fence controller to send electricity through the wire is about $40. Electric fencing is easy to find in home supply stores and other places, but you will probably have the widest selection in a store like Tractor Supply Company. Pigs learn very quickly to stay away from the fence. If you are worried that the electric shock will hurt your pigs, you can try touching the wire yourself. The shock is unpleasant, but it is not harmful. If you put up electric fencing, you should put up two strands of wire. One strand should be about 4 inches off the ground to prevent young pigs from scampering under, and the second strand should be 12 to 16 inches off the ground to keep larger pigs from leaning over the wire. For additional security, you can add a third wire 30 inches off the ground. Make sure you mark the wire every few feet with caution tape to warn the pigs where the wire is. They will likely figure it out for themselves very quickly, but it is still a good idea to do it. You or other visitors to your farm may accidentally walk into the wire without the caution tape as a reminder.

All of these fences will work well for pens and drylots. If you are fencing a larger area, such as a pasture where pigs will be raised, you will probably need to consider the most economical options. Electric wire, barbed wire combined with electric wire, and electrified netting are all possibilities if you are fencing large tracts of land where pigs will graze. Field fencing is one of the most expensive options, and it is hard to install, but this kind of fencing can last up to 50 years. If you are going to fence several acres and you plan to use the land for pasture for a long time, then field fencing may be your best option. You will not need to replace this kind of fence in the near future, and it will do an excellent job of keeping pigs or any other farm animal contained. General purpose field fencing usually starts at $159 for fence that is 47 inches tall and 330 feet long; studded "T" posts for the fence are $3.99 each at the Tractor Supply Company at the time of this writing. As usual, you may be able to find better deals if you buy secondhand. Electrified netting is also expensive, but it is very versatile and can be moved from place to place. It is very useful if you plan to use rotational grazing for your pigs. You can simply move the electrified netting wherever your pigs graze. Most electrified netting in the United States is sold by Premiere1Supplies.com and can be purchased online with free ground delivery. Fifty feet of Pig Quik Fence is $54; 100 feet is $66.

No matter which kind of fencing you decide to use, you should make sure you install it well from top to bottom. Shoats can scamper under fences if they are not at or near ground level, and larger pigs can climb over fences or push against them until they weaken. Part of your daily routine should include casting an eye over your fences to make sure no spots where your pigs can get out exist.

When you install your fences, you will also need to put up a few gates. Give some careful thought to whether gates should swing inward or outward. For instance, if you do not want your pigs to try to rush out of the pen into an open area when you enter, it is best to plan a gate to open inward. If your pen opens into a pasture for the pigs, however, it may be fine to have a gate that swings outward to allow the pigs to have easier access to the larger area. If you have a gate to a barn entrance, you may wish to control access and have a gate that opens inward so you can control how many pigs you allow to enter at one time. Consider the best locations for your gates. If you have adjoining pens, for example, you may wish to place a gate inside the pen as a shortcut instead of having to drive pigs out of one pen and into the other. These may seem like minor matters, but they can make a big difference when you are driving your pigs from one pen to another or carrying out your daily chores.

Hygiene Considerations

Raising your pigs well requires that you give thorough consideration to issues related to hygiene. Your pigs' living quarters will have a direct impact on their health, the quality of meat they produce, and on you or anyone else who works around them. Good planning can make it much easier to maintain hygienic conditions around your pigs, make cleaning the pen easier, and make for a more pleasant work environment on the farm.

What goes in must come out so as you feed your pigs, they must excrete manure. Pigs are naturally clean animals, and they prefer to use one area of their pen or lot as a bathroom. You can take advantage of this natural instinct to keep the pen clean. Pigs move about 12 feet away from a feed or water station before relieving themselves. If a sleeping area is placed at one end of a pen, a feed station is placed on one side, and the water trough is placed opposite on the other side of the pen, then the pigs should walk about 12 feet to the far side of the pen to relieve themselves. Once this area is established as the bathroom spot, your job of cleaning out the pen each day should be decidedly easier because you will only have one place to concentrate your cleaning efforts on. You should also check the sleeping area each day and clean out any wet or dirty places. Put down fresh straw or other bedding to keep the area dry and comfortable for your pigs.

One of the chief reasons for diseases among swine is overcrowding so try to keep the number of pigs you house together down to lower numbers if possible. Hygiene can also be affected when pigs tear up the ground with their hooves. It is also very likely that after it rains, your pigs' pen may turn into a muddy pit. There is not much you can do about mud in your pigs' pen if you raise them on dirt, but you can rotate your pigs to new pens from time to time if you have the space available. This will allow one pen to dry out while your pigs use the new pen. Rotating your pigs to clean pens tends to cut down on the spread of disease and parasites as long as you make sure the new pen has been sanitized since its last use.

You can raise your pigs on concrete flooring. This can be a viable option if you have outdoor concrete pads already on your property where grain has been kept or barns have stood before. Concrete is not an ideal surface for pigs because it is hard on their hooves and legs and not very comfortable if they wish to lie down outdoors. Concrete can also become very hot during the summer months unless there is plenty of shade provided. But, it is easy to keep clean and not subject to disease and parasites in the same way that dirt pens are.

Sand is another option for your pens to keep them cleaner and more hygienic. Pigs enjoy sand, and it is relatively comfortable for their legs and hooves. Sand does not promote the spread of disease or parasites as much as dirt does. Sand can also provide good drainage for your pen so water and urine do not collect.

You can also make your pigs' living situation better and cut down on slipping, sliding, and mud by placing your feeding and watering stations on concrete platforms in the pens. The platforms do not have to be very high. The goal is to simply keep the pigs from sliding around in the mud as they spend time in these very important areas.

Of course, mud will not particularly bother your pigs. While pigs are not the dirty, smelly animals some people imagine, they will make use of a nice mud hole to wallow in during hot weather. A nice coating of mud can protect them from sunburn and keep them cool when it is hot. Mud can also keep insects from biting and stinging. So, a little mud is a good thing. You can also provide your pigs with a nice watering hole. If no natural watering hole or pond is available, pigs will enjoy a child's pool filled with cool water on a hot day.

Remember that ventilation is also important to hygiene if your pigs have an indoor sleeping area or if they are confined part of the time. Good airflow will help reduce disease. If your pigs are housed, you can solve ventilation issues by having one side of their sleeping area open to the outside, having a window in the building, or having a fan in the top of the ceiling.

As mentioned, pigs are naturally clean animals, and they will become stressed and perform poorly if they must live in dirty conditions. If you allow feeders and waterers to become dirty, they will become much more likely to harbor diseases. It may be impossible to keep your pig operation in pristine condition, but you can do everything in your power to keep things clean and tidy. Your pigs will grow better and be happier if you take pride in your work and keep the place looking good.

Food, Water, and Shade

Your pigs have a number of basic requirements to stay happy and healthy. Without food, water, and shade, your pigs cannot grow and be happy. There is a wide variety of watering and feeding methods for you to choose from, depending on how many pigs you will raise, your husbandry methods (the way you raise and care for your livestock), and how much money you plan to invest in your equipment. Some watering and feeding methods will mean much more work for you, but they will cost less. Others methods will cost much more, but you will have much less work to do each day. Before you purchase your feed and water equipment, it is a good idea to see if you can talk to other farmers or visit their farms and see how their equipment is set up. This is especially helpful if you find other farmers who are raising pigs with the same intentions as yourself.

Shade is also very important to pigs, especially if they spend part of their time outdoors. If they are living full time in a drylot or are being pasture raised, then shade is of vital importance. Make sure your pigs have some place where they can get out of the sun. Some breeds are prone to sunburn, and pigs in general do not enjoy extreme heat. If they are too hot or uncomfortable, they will become stressed, and they will not grow or gain weight as predicted. Stress can also lead to aggression and illness so make sure you provide your pigs with comfortable shade.

Waterers

You cannot underestimate the importance of water when raising your pigs. Pigs will need free access to water at all times. You can estimate that each pig will consume about two to three times as much water as feed per day. That means that a 100-pound grower pig that eats 6 pounds of feed per day will consume approximately 12 to 18 pounds of water, or 1.5 to 2.2 gallons. The amount of water your pigs drink will fluctuate slightly depending on the weather and changes in their feed. But, you should continue to have fresh water available to your pigs at all times.

There are a number of different watering methods for pigs, ranging from a simple water dish or tub of water for a single pig to expensive automatic heated waterers. If you are just starting out with raising a pig or two, then it is probably best to start with a smaller investment. You can always upgrade later as you acquire more pigs.

A water dish or tub for your pigs is a perfectly fine way to start out if you are raising one or two pigs. A water dish or tub is made of rubber, plastic, metal, or even concrete. The important thing should be that it is durable so it can stand up to use by your pigs. You will need to refill this water several times per day for your pigs. Pigs do like to turn things over so you can expect them to try to play with the water dish. Pigs will be more likely to turn over lightweight tubs so you can try to use heavier tubs, but they may still play with them. A water trough may be a more practical choice, whether you have one or two pigs or several. You can make a water trough from an old water tank cut in half and placed on concrete blocks or wooden blocks, or you may purchase a new trough from a livestock supply store. Water troughs can be anchored in place, or they may be moved from place to place if you plan to move your pigs to different lots.

Beyond water troughs, watering methods become more complex and more expensive. If you decide you want to use one of these more complex watering systems, you will need to purchase them from a dealer or a feed supply store that carries hog supplies. You may get lucky and find one for sale secondhand.

You may choose to water young pigs by means of nipple drinkers. Nipple drinkers are 1-gallon containers with a metal valve that acts as a nipple on the side. These waterers are good for young pigs, and they keep out dirt and debris, but they have to be anchored securely to keep pigs from dislodging them. The nipple drinkers can gradually be raised higher as the pigs become bigger. Nipple drinkers can be attached to a permanent pipe system so water is constantly available. However, this method of watering is not heated so they are not recommended as an outdoor waterer. Nipple drinkers have to be purchased and are not homemade devices.

A hog watering tank can hold between 35 and 250 gallons of water. These tanks have a "hog drinker," or a trough cut in the side where the pigs can drink. They automatically refill themselves with water, and they can be heated for winter watering outdoors. However, these tanks are very hard to keep clean if you are raising your pigs on dirt flooring. When the pig puts his head into the trough, it is basically washing its face in the drinking water. A hog watering tank does make it easy to give pigs medicine or supplements because you will know how many gallons are in the waterer and can mix appropriately.

The most expensive way to provide your pigs with water is with an automatic heated watering tank. This watering system is beneficial because the water flow is automatic, and this system has a flip lid that pigs learn to use. This prevents dirt and debris from entering the tank. The tanks can also be drained for cleaning, which means they are less work for the farmer.

Feeders

When it comes to feeding your pigs, you can also take several approaches that are similar to your watering options. A feeding dish or tub can work well if you are raising one or two pigs. You will have the same problems with them as with watering dishes, but pigs will play with them and spill their feed, causing wastage. A feeding trough can be used to feed your pigs. If you are feeding several pigs, you will need to use a trough with dividers to keep pigs from being too greedy. Otherwise, a dominant pig may stand in the trough and prevent the other pigs from eating or try to claim all of the feed by pushing the other pigs away from the trough.

Some farmers prefer to feed their pigs by putting feed on the ground for them and scattering it around the pen, away from sleeping and bathroom areas. This method is most often used by small farmers who prefer a more natural approach to raising their pigs because it allows pigs to eat their food off the ground as they would if they were grazing or foraging. If you decide to feed your pigs this way, you should only do so if you are feeding pellets, cubes, or larger feed, as feed of this size is less likely to disappear into the ground. It does not work well with small, finely ground grain because much of the grain will be trampled and go to waste.

Wall feeders are another option for feeding your pigs. Piglets that have not been weaned yet are frequently fed by means of wall feeders in the creep, or the part of the farrowing area where the sow cannot follow them, when they begin having their first meals. This prevents the sow from trying to get their feed and being in competition with them for food. When used in the creep, wall feeders are used until piglets are at least 6 weeks old. Wall feeders can also be used for pigs after they have been weaned if the feeders are raised to the proper height. They are good for feeding six pigs at a time.

Hay racks, such as those used for horses, can be set at a height comfortable for your pigs to use and filled with hay, alfalfa, or other grasses. This is a good option for your pigs when they are indoors. Hay is left in a bale when it is provided to pigs outdoors, or the pigs use the automatic feeders.

If you are raising a large number of pigs, the easiest way to feed your pigs is with a self-feeder, or bulk bin feeder. Self-feeders are very popular, and pigs quickly learn to use them. Self-feeders can hold between 50 pounds and several tons of feed. Self-feeders are expensive — an average self-feeder for three to four pigs will cost $75 to $100, and a small, used bulk bin feeder capable of holding 4 tons of grain may sell for about $1,000 — so you will not want to purchase one of these systems unless you know you intend to continue raising pigs. The trough portion of the feeder has flaps that the pigs open when they want to eat. The farmer can adjust how much feed is available to the pigs. A 40 bushel hog feeder — the Big "O" oscillating feeder with 12 feeding stations — sells for around $1,300 at BarnWorld.com (**www.barnworld.com**). Self-feeders keep feed from being wasted, and they provide the correct rations to pigs, whether feed should be available continually to pigs being grown for market or whether feed should be limited to adult pigs.

Tools

Raising pigs requires that you have certain kinds of equipment, such as waterers, feeders, chutes for loading, pig boards to help move pigs, crates, and snares to assist in leading pigs. You will also have some ordinary farm equipment, such as pitchforks, shovels, wheelbarrows, and the like, for hauling manure. Here are some additional tools you will need when you begin raising your pigs.

Pig boards and snares

Pigs are one of the more difficult animals to move, but there will be times when you need to move a pig. You can carry a small shoat similar to the way you would carry a dog by holding it in your arms and supporting its stomach and chest. The piglet will squeal so be prepared for protests. As the pig gets a little larger, however, it will become much more difficult to handle. You can often move young pigs by using a **pig board** or hurdle. This is a tall board with cut outs or handles in the top so you can grip the board. If you place one board behind a pig and one in front of it to limit the pig's movement options, you can scoot the pig forward in the direction you want it to move. This method works very well if you need to load pigs to walk up or down a ramp to go in or out of a truck, for example. You can also get a pig to go into a dark building like a barn using this method. Most pigs object to entering dark places because of their poor eyesight.

For larger pigs, such as sows, you may need to use a snare to get it to go where you want. A pig's head and snout are not really formed to accommodate a collar or halter. To overcome this problem, pig farmers use a snare. The **snare** is a flexible lead that loops around the pig's upper jaw and snout. Once the snare is in place, the farmer can lead the pig. There is no pain involved. This is often the only way to encourage an 800 pound boar to go where you want it to go.

You can move some pigs by placing a temporary fence around them and scooting them toward your destination. Simply take a shortened section of welded wire or general farm fencing, loop it back upon itself, forming a circle, and place it around the pig. Then, you can walk the pig safely toward your goal, and it will not be able to move away.

Different methods work best with different pigs, depending on their size and degree of cooperation. You will need to practice and find out which methods work best with your pigs.

CASE STUDY

Mikaya Heart
Box 1152
Laytonville, CA 95454
mikayaheart@myfastmail.com
www.mikayaheart.org

Mikaya Heart is a writer, speaker, and life coach. She also spent a number of years raising pigs. "For several years in the 1980s, I sold pork to the some of the best restaurants in the San Francisco Bay Area."

Heart says she raised pigs for about ten years. At the height of her business, she had about 12 sows and 100 feeders. She says that she initially started raising pigs because she liked good pork, and then she continued to raise pigs for the good money. Heart recommends that anyone interested in raising pigs should start with a couple of weaner pigs, about 8 weeks old. "Make sure the males are castrated, or castrate them yourself. It is not difficult to castrate a young pig (younger than 8 weeks) as long as you have a piece of apparatus in which you can hang them by their back legs. Then, the balls are exposed, and you can push them up between your fingers. Make a quick slice through the flesh with your razor so they pop out of the cut. Slice them off, squirt the two cuts with an antiseptic, and let the piglet go."

Heart warns, "NEVER pick up a piglet or make it squeal when its mother is loose and close by. She will kill you. You must have a method of separating the piglets from the mother." She also says that it pays to make friends with your pigs. As she says, it is not easy to herd a pig, and it is hard to force them to go somewhere they do not want to go. You have to trick them, entice them, trap them, get them to trust you, or get them into the habit of going where you want them to go. You can usually do this with food. But, it helps to make friends with them.

Heart also has advice about fencing for pigs. "Pigs really enjoy lots of space and will do best if they can range in woodland. You can keep them behind an electric fence, but make sure the fence has a grounding wire that the pig will touch at the same time as the live wire if the ground is

and on that farm there were...

Pigs!

around the farm

v

happier than a pig in mud!

pigging out!

this little pig went to the market...

raw pork knuckle

pork...the other white meat

raw pigtails

raw pork shank

smoked pork shank

salted pork fat with skin

this little pig stayed home!

too dry to let a current flow. I found I had to have a live wire a few inches above the ground just inside a regular fence, which acted as the ground. They would catch it with their noses when they were digging. But, they would also tend to throw dirt over it so I had to check it regularly."

She also said, "Pigs don't like being handled by strangers so if you can give them any shots they need yourself, that is by far the best option. Occasionally, they will need antibiotics if they get an infection. Best if you can learn to take their temperature, diagnose that yourself, and then buy your own antibiotics. If you are breeding pigs, make sure their farrowing pens are really warm and clean, as the piglets are very prone to disease for the first few days."

General farm equipment

In working with pigs, you will need some general farm equipment to handle things like manure removal. The usual way to handle manure is with a pitchfork or shovel. You can then toss it into a wheelbarrow to take it where you intend to store it or place it for composting. A pitchfork is the more effective tool if you are working indoors and removing bedding because it allows you to tease out dirty bedding and leave behind the clean bedding, which saves money in the long run. A shovel may be more practical in the pigs' pen, especially if you have several pigs. You can purchase implements such as these at any farm supply store in your area. You can always buy them secondhand if you wish to save money. If you do purchase pitchforks or shovels secondhand, you should make sure to scrub them thoroughly before using them so you will not transfer any parasites from the previous owner's property to yours.

It is very useful to have a pickup truck if you are raising pigs. Hauling grain and other supplies, and even hauling pigs at times, will be much easier with a dependable truck. Depending on the size of your farm, you may want to think about getting a small tractor. If you have crops, you will probably already have a tractor and tiller for tilling the ground and planting seeds. If you have pasture, you can attach a manure spreader to a tractor and easily spread the manure over your fields to fertilize them.

If you have a larger property or you intend to raise a large number of pigs, you may wish to consider purchasing a manure spreader. You will have a lot of manure to contend with on a regular basis, and a manure spreader would allow you to regularly spread the manure over your pastures as fertilizer. Cart-type manure spreaders that attach to tractors and hold 25 bushels of manure can be found for around $900 and will work very well for hobby farmers.

From time to time, you may need equipment for putting up fences, such as post hole diggers or other specialized tools for certain tasks. You may consider borrowing these tools from other farmers instead of buying something for one-time use. You may also be able to rent some items for one-time use.

Things you may need to use on a short-term basis include:

- Post hole diggers to dig holes for putting up fence posts
- Fence stretcher for pulling fence wire between posts
- Tiller for tilling soil and compost
- Forklift for lifting bales of hay and multiple bags of heavy grain
- Backhoe for occasional use when you may need to dig a trench or bury a large dead animal
- Bush hog for cutting pasture

⑤ Building supplies if you intend to build a permanent structure for your pigs, a smokehouse, or any shelters for your pigs in your pastures

Regional Considerations

Where you live will help you make some of your housing and other decisions. The Midwest has long been the hog capital of the United States and that remains true today. Most large commercial pork producers are located in this area. As a result, even small farmers in the Midwest will benefit from some of the breeds and services available in this region that may not be readily available elsewhere.

In terms of housing, you will need to consider the weather where you live. If you live in the South or Southwest, shade and the ability to keep you pigs cool during the summer will be very important considerations. In some southern states, it is not unusual for farmers to turn on misters or sprinklers so their pigs can cool down when it is hot. In northern areas, you will have to pay special attention to winter temperatures and the possibility of providing extra heat for some breeds of pigs, especially during and after farrowing. If you live in the Pacific Northwest, you will have to find ways to cope with raising pigs in a damp climate.

Different areas may also have different approaches to manure disposal, pig waste in the ground water, and parasite and disease control. You should talk to your county extension services agent to find out the recommended procedures in your area. If you intend to pasture raise your pigs, you will probably want to find out what kind of grasses and crops grow best in your area, in case you plan to plant them for your pigs. Again, your county extension services agent can assist you with information about grasses and crops. The county extension service can also perform a soil analysis to find out the composition of your soil. Knowing what is in your soil will also help you know what nutrients your pigs are getting when they graze on it.

It will help you to talk to some of your local pig raisers before you get started so you can find out how farmers in your area handle issues in your region. You can also talk to your county extension services agent about regional matters relating to growing pigs. He or she can provide you with a wealth of information about the soil in your area, crops, pig raising, pig housing, and other materials you will need. Your veterinarian is another great resource. He or she can answer questions for you about disease and parasite control in your area and what vaccinations will be most useful to your pigs.

Summary

In this chapter, you have learned about some of the different structures you can use to house pigs, from permanent structures and Quonset huts to hoop shelters and farrowing huts. In the next chapter, you will learn more about one of the most important parts of raising pigs: their feed. Although pigs will eat most things you give them, feeding your pigs can be a complex subject if you are to achieve optimum weight gain efficiency, which is the best weight gain for the amount of feed you are feeding them. The right feed can make all the difference when it comes to raising pigs you can be proud to take to market.

CHAPTER 6:

Eating Like a Pig

With your pigs purchased and your pig housing resolved, it is time to consider what to feed your pigs. Although pigs are not too picky about their food, feeding them the right feed can lead to better and faster weight gain, which can mean better profits for you. Every pig raiser wants to see happy, thriving pigs, and that depends on knowing something about feed.

Everyone has heard the expression "eating like a pig." As it turns out, there is a good reason for that saying. Pigs do eat like pigs. That is to say, they like their food — a lot. A pig's appetite and its ability to gain weight quickly compared to other farm animals are traits that have made it dear to the hearts of farmers. At one time, pigs were known as "the mortgage lifters" because of their excellent return on investment to the farmer and their ability to help pay the mortgage. Just a little feed and some scraps tossed your pig's way for a few months, and it can present you with a freezer full of delicious pork. If you grow an extra pig, you can sell the second pig at market and more than make up your feed and other costs.

If you are like most first-time pig owners, you may be filled with some concerns about feeding and nutrition. Fortunately, pigs are very efficient animals, and they can thrive on different kinds of diets. For instance, if you pasture raise your pigs, they can get about 60 percent of their nutritional needs from the field. You will need to provide them without about 40 percent of their dietary needs from feed and supplements.

For pigs raised in pens, you can expect to provide most of their nutrition from grain in some form, along with as many table scraps as you can provide. Many small farmers make a habit of collecting vegetables, bread, and other things from their kitchens each night, adding some milk to the food remains, and giving this mixture to their pigs. This is a modern form of slop that has been given to pigs for centuries. People once kept slop pails beside their back doors the way they have a garbage disposal unit now.

If you have a garden on your property, your pigs will welcome anything you wish to toss their way. Some people suggest that pigs should avoid potatoes and other plants of the nightshade family. An occasional potato or the peelings from a potato probably will not hurt your pigs, but it is best to avoid them. Young potatoes often have a tint of green on them and are high in the toxin solanine, which is a glycoalkaloid that can be hard for your pigs to digest. Eating a few may not harm your pigs, but it is best not to feed them a large amount because they can cause diarrhea and other gastrointestinal problems or in extreme cases, death. Otherwise, you should feel free to offer your pigs most things from your garden. You can also check with grocery stores to see if they have leftover produce. You should also ask local bakeries if they have day-old desserts or breads they would let you have for your pigs. Pigs enjoy variety as much as you do so as long as something is not toxic to them, there is no reason you should not include it in their diet.

If you are trying to feed your pigs an organic diet, then you will probably want to avoid feeding them meat scraps, unless your meat scraps are certified organic. If you are pasture raising your pigs, they will likely eat small rodents from time to time and benefit from the protein. However, some people do not like the idea of intentionally including meat in their pigs' diets. Pigs love eggs, and it is fine to feed them eggs, shells and all. If you have hens on your farm, you will have a constant source of eggs for your pigs. But, you should discourage your hens from laying their eggs near the pigs. Pigs will quickly learn to follow the hens and eat their eggs if they can get to them.

If your pigs are allowed to forage, then they can find many tasty things to eat on their own. Pigs that eat acorns and chestnuts will have particularly flavorful pork. But, you should try to keep your pigs away from areas where black walnuts fall because these can be toxic to your pigs. The American Society for the Prevention of Cruelty to Animals® (ASPCA) has a good website with photos of more than 400 toxic plants (**www.aspca.org/pet-care/poison-control/plants/**). Not all of these plants are toxic to pigs and not all parts of a plant are toxic in some cases. If you have questions about the plants growing on your property, you should consult your county extension services agent or someone from your local university's agriculture or biology departments. They will be able to identify plants for you.

The following plants are known to be poisonous to pigs so you should learn to identify them and keep pigs away from them or remove them from your property:

- Black-eyed Susans
- Jimson weed

- ⑤ Lambs quarters
- ⑤ Nightshade
- ⑤ Pigweed
- ⑤ Pokeberry
- ⑤ Spotted water hemlock
- ⑤ Two-leaf cockleburrs

Grains

Life is not all slop and leftovers for pigs. Most of your pig's diet will probably come in the form of grain. Most grains have high concentrations of energy so they help your pig gain weight more quickly. Grains form the basis of most feeds you can buy for your pig. You can grow your own grains as crops, which is an economical way to feed your pigs, you can buy grain from a local feed mill, you can purchase commercial feeds that rely heavily on grains, or you can have custom grain mixes made up for your pigs by your local feed mill.

There are five basic grains in common use for feeding pigs: corn, wheat, milo (sorghum), barley, and oats. Corn is the most common of these grains. When the other grains are used, such as wheat or barley, they are not used as a full grain feed ration, or as the sole form of grain that is fed to the pig. Instead, they only make up part of the grain ration that is fed to the pig. This is because they are less palatable in some cases and have fewer of the nutrients that pigs require. When you feed your pigs these grains, they will need more supplements. Corn can be fed as a full grain feed ration by itself, which explains why it is so popular as a feed for pigs.

Corn is high in carbohydrates and low in fiber. It is an especially good feed for **finishing** hogs, which when you are trying to put the last pounds on a hog before going to market. However, corn lacks amino acids and is low in protein. It has a protein content of 6 to 9 percent. Pigs need a minimum of 13 percent protein, depending on their growth stage. This means that your pigs cannot live on corn by itself. You will need to supplement your pigs with protein and vitamins and minerals if you feed them corn.

Protein

Your pigs must have good sources of protein in order to have the amino acids they need to thrive because their bodies cannot make amino acids on their own. **Amino acids** are necessary for muscle building, for gestation, for lactation, and for growth.

Because protein is essential for growth, pigs need different levels of protein at different times in their lives.

Pig Protein Requirements	
Age	Protein %
Nursing/weaning	18-20
Growers (50-125 lbs)	15-16
Finishing (125-250 lbs)	13-14
Young gilts and boars	15-16
Adult sows and boars	13-14

As you can see, young pigs need the most protein because this is the time of their fastest growth. As pigs get older, their growth slows and their protein requirements taper off slightly.

There are a number of good protein supplements commonly added to a pig's feed. Soybean meal is a popular protein supplement. Soybeans can be easily grown in case you are interested in growing your own crop. You should not allow pigs to eat raw soybeans, though. They contain trypsin inhibitors, which can prevent the young pig's body from absorbing protein properly and inhibit growth. Soybean meal does not contain this ingredient. Most of the trypsin inhibitors in soybeans are neutralized by heat during processing to make soybean meal. Although soybean meal is economical, pigs find it very tasty, and it is a good source of protein, it still lacks vitamins and minerals. If you use soybean meal as your protein source, you can use alfalfa meal as a good source of vitamins and minerals for your pigs. Mix the soybean meal, alfalfa meal, and corn together for a good, nutritious feed for your pigs.

Other plant sources of protein include wheat, bran, alfalfa meal, distiller's grains, brewer's products, corn gluten meal, and hominy feed. Depending on the grain and protein supplement you choose for your pigs, you will need to find the right mixture of vitamins and minerals to add to your pigs' diet. For example, if you use alfalfa meal as a protein source, it is already a good source of vitamins and minerals, meaning that you will have to use fewer vitamin and mineral additives. The downside of using some of these other plant protein sources is that they are more costly than soybean meal. For example, at one feed store, a 100-pound bag of soybean meal was $19; a 50-pound bag of alfalfa meal was $12; and a 50-pound bag of corn gluten meal was $24. You will also need to consider the fact that soybean meal has approximately 38 percent protein, compared to 16 to 18 percent protein for alfalfa meal, and 40 percent protein for corn gluten meal. So, choosing the best protein supplement is not always easy.

CASE STUDY ™

Ryan P. O'Neil Sr.
3 Tree Farms
Owner Operator
2142 Thiel Road N.
Collins, NY 14111
Ryan@3treefarms.com
www.3treefarms.com
716-392-8129

CLASSIFIED CASE STUDIES
directly from the experts

Ryan O'Neil of 3 Tree Farms in New York says that he raises happy, healthy pasture hogs. He has six sows and one boar. "We eat healthy and with the knowledge that our animals are humanly raised and bred. We also offer this same peace of mind and health benefits to our customers. Hogs take a bit of organization, preparation, and planning. All are great skills to learn, along with the end product being so beneficial. It is great for families and namely children to develop these skills, and there is the added bonus of achieving success at something so productive."

According to O'Neil, young feeder hogs would be a great way to start raising pigs. "Many first timers start with spring feeders and raise for the table in the fall. Then, you don't end up overinvested in a project. If your passion isn't there, you have a freezer full [of meat] and the knowledge that you may never try again. However, if you become passionate and want to engage your time further you can. If they want to take the next step, they can move on to keeping a sow for breeding or purchase a bred-back sow in the following spring."

O'Neil's pigs are pasture raised, and he rotates them. "I have pens I have designed, and I keep the hogs in pastures with Osage orange plantings as fencing. We feed them hog mash, corn, and apples. We plant our pastures with turnips and rotate the hogs through on a 17-day cycle. We use three pastures and that gives time for the first pasture to [regrow] before the hogs get back to it."

O'Neil is careful about what his pigs eat. "[We feed] mash, apples, corn ears is a favorite, pumpkins, squash, and whey from goat cheese making. We pasture turnips. No scraps and junk. We don't have slop hogs. We compost our junk and feed our hogs healthy."

O'Neil tries to raise his hogs naturally. "I am humane to my hogs. I do not give any meds, steroids, or any other unnatural preventatives. However, like my children, if they're sick I treat them."

Grain analysis

Before figuring how much protein you need to use as a supplement, it is always a good idea to have your grain analyzed. A grain analysis can tell you how much protein is actually in the grain you use so you will know how much protein supplement you need to use. If you use an independent feed supplier, you can have the supplier perform an analysis of your feed. There will be a sample preparation fee that is normally less than $30 for a basic analysis that examines your minerals and protein. Or, you may send your feed to an independent laboratory for analysis, such as a university agricultural school. Otherwise, you may feed your pigs too much protein. Pigs and other mammals cannot use protein past an optimum point. Once the body utilizes as much protein as it needs, the excess protein is either excreted through the urine, which can contaminate ground water, or the excess protein will cause the liver and pancreas to enlarge.

Vitamins and Minerals

Unlike cows and horses that can take minerals from a mineral block in their pasture or stalls, pigs cannot use mineral blocks. Some of the ingredients in mineral blocks are toxic to pigs so they should be avoided. Vitamins and minerals normally come in premixes to be mixed into grains or feeds, unless they are supplied in the form of a specific supplement, such as alfalfa meal. But, premixes can be expensive and are best purchased in bulk. For one brand, a 60-pound bag of starter premix was $37.50 with a ten bag minimum order if purchased online and delivered. Premixes can be purchased at many feed stores where livestock feed is sold. Vitamins and minerals need to be given to your pigs in the right formulation because too much or too little of certain ingredients can lead to health problems. For example, selenium is toxic even in low amounts, though a lack of selenium can cause sudden death in rapidly growing young pigs. It must be fed at just the right amount — 3 parts per million for pigs. Less salt (sodium) is good for finishing and growing pigs. But, an increase in salt in the sow's diet from .025 percent to .5 percent can increase litter size by half pig per litter. The vitamins and minerals your pigs require will also fluctuate at different stages of their development. Your pigs can rely on some vitamins and minerals from the soil if they graze, and they will get some vitamin D from the sun if they are outdoors. However, pigs confined indoors will rely completely on vitamin and mineral supplements.

Other supplements and additives

Along with vitamins and minerals and protein supplements, your pigs may benefit from other additives to their feed. Meat and bone meal, fish meal, and tankage are all animal-based sources or protein that can add high quality protein and/or fat to your pigs' diets. Animal-based sources of protein are normally more expensive than the plant-based sources, but they have high amounts of protein in small portions so you can use less of them, saving money. Meat and bone meal and fish meal are leftover byproducts from slaughter and processing. **Tankage** is a liquid product from rendered (cooked down) animal carcasses. When it comes to palatability, pigs seem to prefer the taste of soybean meal to other protein supplements. Meat and bone meal, fish meal, and tankage are good sources of phosphorus and calcium. They can be used for up to half the necessary protein in your pigs' diets.

You can purchase meat and bone meal and fish meal from your feed supplier who can add them to your feed mix. You can also purchase fish meal from your feed supplier, from feed stores, or even in garden centers. A typical price is $13 for a 2-pound box, although you may need to purchase it in much larger quantities.

Pigs are very fond of milk products, and these products can supply many of the nutrients missing from grains. Milk products are easy to digest, and they are especially good for young, growing pigs. Dried whey and dried skim milk are added to the starter mix diets of young weaned pigs in order to improve growth performance. They can be fed until the pigs weigh about 30 pounds. Lactose is milk sugar, and dried whey contains 70 percent lactose; dried skim milk is made up of 50 percent lactose. Whey contains lactose, lactoglobulin protein, lactalbumin, water-soluble vitamins, and minerals. Dried skim milk contains a similar content but has less lactose and more milk proteins. Dried skim milk costs more than whey and is reserved for younger pigs. If you have a cow or two on your property, you may easily be able to spare some milk for your pigs. You can still give milk to your adult pigs, though they will not benefit from the milk as much as young pigs. It is fine to give your pigs milk straight from the cow. You can also give your pigs cheese or yogurt if you like. Molasses is often added to a pig's grain mix both to improve palatability and to help keep dust down. Good grain should not be dirty or dusty. However, adding several finely ground supplements may make the grain ration become dusty and powdery. The molasses will solve this problem. Molasses is high in iron, but it is also high in calcium, which prevents the pig's body from absorbing the iron.

Commercial Feeds

If you are starting to think that feeding your pigs is too complicated, there are very good commercial swine feeds available at feed stores. They come in many different formulas: for starting weaning pigs, for shoats, for growers, for finishing hogs, for pregnant sows, and for breeding. Some areas of the country have more choices in feed available to them than others, but commercial swine feed is available wherever livestock feed is sold.

There are many popular pig feed manufacturers, such as Purina Mills®, Lindner, Nutrena®, Tractor Supply Company, Starmaster, Blue Seal®, and many others. You can ask other pig farmers in your area or online for recommendations.

The Midwest has more hog farmers than other parts of the country, and it is easier to find a wide variety of feed there. But, there are pig farmers in all parts of the United States, and swine feed is sold in most rural areas. If you can locate a good feed store, they will likely carry some of these brands, or they may be able to order them for you. If you do not live near a feed store, you may be able to find a distributor who sells online or over the telephone if you are willing to buy in bulk.

Companies such as Purina Mills (**www.swine.purinamills.com/**) raise pigs and carry out research about nutrition and diets. Land O'Lakes Purina Feed and the Longview Animal Nutrition Center work with veterinarians, nutritionists, and managers to do their research and develop products for pig producers. They maintain more than 200 sows that farrow in 40 crates, producing 40 litters every five weeks. Most of the pigs are used in research for nursing, growing, and finishing pigs. The facility has eight nursery rooms and can hold a capacity of 800 pigs. The finishing facility has room for 952 pigs, with pens designed for groups of six to eight pigs. Purina can carry out nutrition research using these pigs. They also have an online library of information about swine nutrition. It contains information about new products, how to maximize litter weaning weights, weaning age, immunity and gastrointestinal function at weaning, EPA regulations for commercial swine producers, and more.

Organic Feed

According to the latest reported figures from the National Agricultural Statistics Service of the USDA in 2010, for the year 2008, there were 8,940 hogs and pigs certified organic in December 2008 on 258 farms in the United States. In order to be certified organic, pigs must be raised organically from the last third of the gestation period until butchering. Antibiotics may not be used, which includes antibiotics in feed. Growth hormone stimulants may not be used, and artificial ingredients and preservatives are kept to a minimum.

If you plan to raise your pigs organically, then you will need to start with shoats that have come from a sow that was raised organically from the last third of her gestation period. You can begin an organic program with grower pigs that have not been raised organically and change over to organic, but you will not be able to sell their pork as organic. In this situation, the piglets will need to be weaned onto an organic starter feed. You should continue to feed them an organic diet by feeding them organic grain and supplements raised on your own farm or purchased from a local organic farm. Sources for organic grain can be harder to find than commercial feeds, but they are available. Some feed mills offer organic grains, and you can have custom mixes made to your specifications. You can contact individual organic farmers in your area to see if they sell organic grain. Talk to your county extension services agent to find out about organic farmers near you. He or she may also be able to suggest other sources for organic grain. You can find national sources of organic grain through ATTRA, the National Sustainable Agriculture Information Service site at **http://attra.ncat.org/attra-pub/livestock_feed/**. They have listings for organic livestock feed suppliers throughout the country. There are also commercial organic feeds available that do not contain antibiotics or other preservatives. You will need to check to see if these feeds contain genetically modified corn or other grains. Modesto Milling (**www.modestomilling.com/**), the Buckwheat Growers Association of Minnesota (**www.buckwheatgrowers.com/**), and Ranch-Way Feeds (**www.ranch-way.com/**) are a few companies that create organic feeds for pig farmers.

As usual, it will help to talk to other farmers who raise pigs this way, even if they do not live in your area. They can offer you tips and advice to find the best organic feed for your pigs. You can also check with your county extension service agent about organic resources in your area. You can find information about the county extension service online by searching for your state's land-grant university, such as the University of Tennessee, and "county extension service." Then, look for information regarding your county.

Custom Mixes

If you raise your own grain or if there are farmers in your area who raise grain, you have the option of going to your local feed mill and requesting a custom feed mix for your pigs. This can be a good choice if you are not happy with the commercial choices available to you. This approach is the most cost effective if you have a large number of pigs and you can purchase several tons of custom mixed grain, but your local feed mill can also custom mix grain for you and sell it to you by the 50-pound bag.

If you have your grain custom mixed, you can have your protein supplements and your vitamins and minerals added at the same time the grain is milled, which means less work for you because you will not have to mix up each ration for your pigs. Custom mixes can be great labor savers if you have more than a few pigs that eat the same diet. If you live outside the Midwest "Hog Belt" and you have fewer kinds of commercial feed available, then having a custom mix created can allow you to make sure your pigs get the exact ingredients you want them to have. You may be able to buy your grain from local farmers so you will know where the ingredients came from and how they were grown. You will also save money if you buy your feed locally from the feed mill. Your feed can cost about half the price of commercial feed if you buy in 1,000 pound amounts from the feed mill and bring it home yourself in a pickup truck.

Storage

The subject of feed inevitably leads to the issue of storage. There is almost always a cost savings to buying grain and supplements in bulk, but buying in bulk makes storage more difficult. Just where do you put 3 tons of grain? Bulk bins are the best place to store large amounts of grain, and they have self-feeders in the bottom so pigs can feed themselves. At the moment, you may not think you would need a ton of grain, but keep in mind that one pig will eat about 7 pounds of grain per day. That is about 50 pounds per week, or 200 pounds per month. Multiply that by your number of pigs. If you have five pigs, you will easily use 1,000 pounds of feed in a month. So, it is not unusual to buy 2,000 pounds or more in one purchase.

Feed mills can put grain into a large bag for you that will hold 1,000 pounds. They can place this bag in the bed of a pickup truck, and you can bring it home. A tractor can move the bag to a bulk bin or you can dip the grain out in just a few minutes by using a 5-gallon bucket.

You do not have to purchase tons of grain at one time. You should not purchase more than a month or two worth of grain at one time to prevent the grain from becoming dusty or moldy. You should never feed your pigs moldy grain. Moldy grain can be a sign that the grain is bad. It may be harboring aflatoxins, which can make your pigs seriously ill or lead to death. Store your feed where it will stay dry and where vermin will not get to it. Feed will always attract rodents so take care that you use containers that are rat-proof. For instance, you can store a 50-pound bag of feed in a metal trash container with a lid that fits securely. This will keep the feed dry and keep rats out. Make sure you never place feed containers where your pigs can get to them or they will help themselves.

Summary

This chapter explored some of the different things you can feed your pig, especially grains and supplements. This chapter also provided information about different protein sources and supplements. Armed with this information, you should be able to make good decisions about what to feed your pigs.

Next, you will explore the health of your pigs: how to keep them healthy and how to recognize when they might be ill. Losing a piglet or an entire herd can be devastating so take every precaution to keep your pigs healthy.

CHAPTER 7

Hog Health

With good care, good feed, and good housing practices, your pigs should stay in good health if you have purchased them from disease-free sources. However, pigs can succumb to disease if one of the following situations occurs on your farm:

⊚ If your pigs' needs are not being met in some way;

⊚ If they are not getting the proper nutrients;

⊚ If you are overcrowding animals;

⊚ If you are not practicing good husbandry and you are failing to clean out the pigs' pen or keeping things in poor condition; or

⊚ If the pigs are being stressed in some other way.

Recognizing Symptoms

Part of your daily care of your pigs should include observing them to make sure they are in good health. You can often do this when your pigs are eating. If there is a pig that is not hungry or that stands by itself or if there is a pig that is listless or that seems different from the day before, you should notice it and take it seriously. It is highly unusual for a pig to miss a meal. Pigs are social animals, and if a pig is by itself, something may be wrong. Pigs should not be listless or uninvolved in the things that are going on. The more you watch your herd, the more you will get a good feel for how your pigs behave when they are healthy and feeling good, and the sooner you can recognize when something is wrong with them.

Loss of appetite is an obvious sign of illness in most pigs. Pigs love to eat, and if you have a pig that refuses to eat, you should find out what is wrong with it right away. Very young pigs cannot miss many meals without becoming much sicker.

Your pig supplies should include a basic first aid kit for your pigs. This kit should contain many of the same items you would include in a first aid kit for your family, such as triple antibiotic ointment, iodine, alcohol, styptic powder, bismuth subsalicylate, and similar treatments. It should also contain a rectal thermometer. You can use a rectal thermometer to take your pig's temperature if you are concerned that the pig may have a temperature. The normal temperature for a pig is 102.5 degrees, and anything more than 103 degrees can be a fever. If your pig does have a temperature, it can help you identify if your pig is ill and which disease it may have.

Diarrhea or scours is a frequent symptom of diseases in pigs. If your pig has diarrhea, it may become dangerous very quickly, particularly to very young animals because they can become dehydrated. You can add electrolytes to your pig's water and encourage it to drink to help it. You can purchase powdered electrolytes to mix in your pig's water, or you can purchase a substance, such as Pedialyte® liquid. Try to get your pig to drink enough to replace the fluids it has lost, which may be quite a lot of fluid. In a pinch, you can use sports energy drinks with electrolytes. Try to quickly determine what is causing your pig's diarrhea and call your veterinarian for help in case it is something serious.

Diseases

Pigs are subject to at least 140 diseases. In order to raise pigs, you should purchase a good veterinary manual to help you identify pig diseases and ailments. You can also visit The PigSite for more detailed information about pig diseases at **www.thepigsite.com/diseaseinfo/**. Consultants for The PigSite include several veterinarians, pork industry consultants, and geneticists. They have an extensive section on pig diseases and pig health. It is an international site with very good information about all aspects of raising pigs.

Here are some of the most common pig diseases and their symptoms:

Atrophic rhinitis

Atrophic rhinitis is an illness that infects young pigs and leads to physical deformity. It is much more difficult to detect in early stages on lighter-skinned piglets. The cause is presumed to be germs that enter the mucus membranes of the nose. Nose to nose contact infects one animal to another, until the entire litter is sick. Symptoms do not show up until piglets are 3 weeks old. There is normally no fever, but the pigs have sneezing and a discharge from the eyes, along with inflamed mucous membranes. The most noticeable symptom is an acute irritation of the nose, which makes the pigs rub their noses against anything they can find. The disease will eventually affect the bones of the snout and deform them. The face will gradually become misshapen, and the bones around the face and nose may disintegrate. The snout may become very noticeably curved along the side. Young pigs are very prone to pneumonia with rhinitis. Up to 30 percent of pigs with atrophic rhinitis may die. Testing is available to identify carriers of atrophic rhinitis, and it is recommended that new animals be tested before they join your herd to prevent them from passing this disease along to your pigs.

Some treatment protocols may help the surviving herd members. The breeding herd should be vaccinated against the disease. It may take up to four months for immunity to develop. Use sulfa drugs to medicate sows through their feed from the time they enter the farrowing site until they wean their piglets. Inject piglets with amoxicillin on days three, ten, and 15 while they are nursing, and inject them again at weaning time with a long-lasting antibiotic. Continue this method of vaccination for at least two months until all of your sows have been vaccinated. Give sows a booster vaccine two to three weeks prior to each subsequent farrowing.

This sickness can carry on through quite a few litters if it is left untreated. Inoculating farrowing sows will help prevent it, but rhinitis management has to be done on a continuous basis.

Brucellosis

Brucellosis is a devastating disease among pigs and other livestock, and the illness affects reproduction. The death rate among mature boars and sows is very low, but the disease manifests itself in aborted litters, arthritis in sows, and inflamed testicles in the boar. Brucellosis can be passed through feed contaminated with urine, manure, or other discharges from affected pigs. It can also be passed through shared water. Brucellosis is also sexually transmitted so if you breed your sow to an outside boar, you will need to ask for proof that the boar has been recently tested for brucellosis. The owner of the boar may ask the same for your sow. If an infected sow does have a litter, she will pass the disease along to the nursing pigs in her milk. The only treatment for brucellosis is to remove animals from the herd and to disinfect the entire area. If you are dealing with an outbreak affecting only a few animals, you may be allowed to test them with blood tests and remove only the animals that test positive for brucellosis. In most cases, antibiotics are not very effective. If the entire herd is affected, the only reliable way to prevent the spread of the disease is to destroy the herd. It is required by law in some areas because it is a matter of public health. Many states require certification that pigs are brucellosis-free before allowing them to be shipped into their area. You can check with your veterinarian or your county extension service agent to see if brucellosis testing is required in your area before shipping. A brucellosis test should be covered on a health certificate when purchasing new pigs.

Hog cholera/African swine fever

At one time, hog cholera, or classical swine fever, was one of the most deadly of all swine diseases. The symptoms include a high fever, lack of appetite, diarrhea, and coma leading to death. It is extremely contagious. African swine fever resembles hog cholera, and it is present in the United States. Any pig that displays a high fever should be isolated to prevent contagion. There is no treatment for hog cholera.

African swine fever has symptoms very similar to hog cholera. Lab tests are needed to be able to distinguish African swine fever from hog cholera. Otherwise, the virus exhibits the same sudden onset. Pigs may be found suddenly dead. Pigs may have high fever, lack of appetite, and seem listless. The pig may show red or blue areas on its chest or stomach and on its extremities. There is diarrhea, problems breathing, vomiting, and miscarriage in pregnant sows. When these symptoms occur, virtually all pigs exhibiting them will die within one week. If a pig does recover, it will be a carrier for the virus throughout its life.

African swine fever is spread from one pig to another. There is no effective treatment or vaccine at this time. When African swine fever appears in a herd, it is essential to slaughter the herd immediately before the virus can spread to other farms. Otherwise, the entire pig population is at risk. Everything on the farm must be cleaned and disinfected or discarded. A quarantine or isolation zone may be set up around the area to see if other farms may show signs of the illness.

Parvovirus

Parvovirus is believed to be widely present in pig herds. It does not have noticeable outward symptoms or cause death in mature animals. Instead, parvovirus results in reproductive problems. Parvovirus infection can lead to miscarriages in the sow between day 12 and 20 after breeding and a delay coming in season for rebreeding. Once infected, the sow can continue to lose litters and be slow to come in season, or to come in heat again, for rebreeding. There is no treatment for pigs with parvovirus.

There are ways to prevent parvovirus from occurring in your herd. There is a vaccination against parvovirus. Buying disease-free pigs from established farmers who have no reproductive problems is another safeguard. You can also maintain a closed herd, meaning that you do not bring in new pigs that could pass along diseases to your pigs. If you do have new pigs, it is best to pen your current gilts and dry — unbred — sows near them a few weeks before breeding. This way, they can come in contact with any potential viruses or germs prior to breeding and develop antibodies. Later, after breeding, it would be more dangerous to incubate any new virus because it would cause a loss of the embryos. The gilts and sows will then be able to pass their antibodies to parvovirus or any other viruses along to their piglets. You can then have the piglets vaccinated for parvovirus when they are a few weeks old. Keeping your gilts and sows up to date on their vaccinations is always recommended.

Porcine Reproductive and Respiratory Syndrome (PRRS)

Porcine Reproductive and Respiratory Syndrome, or PRRS, first became noticeable among swine in the United States in the mid-1980s. It was identified as a virus and named in 1991. PRRS has a host of symptoms and is sometimes confused with pseudorabies. PRRS affects the pig's lungs, which leads to reduced oxygen levels. There is a high fever, loss of appetite, coughing, and breathing problems. There can also be reproductive problems, such as late-term abortions for pregnant sows. At the moment, there is no treatment for PRRS, but veterinarians may treat the pig with antibiotics to prevent or lessen secondary infections. There is a vaccine for PRRS, but it seems to be only moderately successful against the virus at the present time. There are research efforts to make a more effective vaccine.

Pseudorabies (mad itch)

Pseudorabies is a deadly disease for piglets. It can also be fatal for mature pigs. Symptoms for piglets come on very rapidly and can lead to death in 24 hours. The symptoms include paralysis, fever, coma, and death. Among adult pigs, the symptoms include the telltale "mad itch," which can make the pig rub itself raw before dying. Pseudorabies is very contagious and is caused by the herpes virus. Most states require testing for this disease because it easily spreads to other farm animals and causes death. When it initially shows up on a farm, the symptoms are manifested in high numbers of miscarriages and weak pigs.

If you experience a high number of newborn fatalities, do not automatically assume it is due to a poor sow. Sudden onsets of illness like this can cause mortality rates to skyrocket. Consult with your veterinarian, and establish a plan for protection of your herd, no matter how large or small.

Sick pigs show signs of pneumonia, high fever, disorientation, and convulsions. Older pigs that are weaned are more prone to pneumonia and fevers. There is no medication for treatment, although the pneumonia symptoms can be treated with some forms of antibiotics.

There is a vaccination for pseudorabies, but it is only available in states where pseudorabies is considered widespread. The vaccine does not prevent pseudorabies; it only prevents the symptoms from manifesting. The United States declared pseudorabies to be eradicated from the commercial pig sector in 2004, but the disease is still found among the feral pig population. Feral pigs are found in most states so pig farmers must take precautions to keep their herds away from them. Pseudorabies is a legally reportable disease: If it appears in your herd, you must report it to your veterinarian and health officials. The disease may be transmitted to other animals, and they may display symptoms similar to rabies, though pseudorabies is not related to rabies. Incinerating the carcass may be necessary after a pig or herd dies from pseudorabies.

Pseudorabies has earned the nickname of mad itch because of the uncontrollable itching it causes. All infected animals have to be separated out and destroyed to protect the rest of the herd. Avoid purchasing pigs that exhibit any of these signs and symptoms.

Swine dysentery

Swine dysentery may be identified by diarrhea, particularly bloody diarrhea. It is most common in the Midwest among larger pork producers. The disease affects the large intestine in pigs. It is acute, coming on quickly and rapidly escalating. It is also extremely fatal. So far, there has not been much success with a vaccine for swine dysentery. The disease is traceable to sales barns or to people who have tracked manure from an infected farm on their shoes. It can affect any pig, but it most often affects young growing pigs. It is normally spread by pigs picking up or eating manure from a sick pig or a pig carrying the disease. Symptoms include loss of appetite, fever, and diarrhea, which may become bloody. If your feeder pigs contract swine dysentery, it is not unusual to lose up to 20 percent of your pigs. The pigs that survive will likely be stunted in growth and may do poorly. There are currently no specific cures for swine dysentery. The best way to stop it is to keep it from coming to your farm. Practice good hygiene, quarantine new pigs, and ask visitors to wear plastic over their shoes to prevent the spread of disease.

Swine erysipelas

Swine erysipelas is a bacterial disease that occurs in young pigs between 3 and 12 months old. Just because it is a bacterial disease does not mean that it is not serious. Swine erysipelas can be fatal. The erysipelas bacteria live in the soil on the farm, and once the bacteria establish, they can be very hard to eradicate. Infected pigs continually reinfect the soil with bacteria in their urine and manure. Symptoms of swine erysipelas can be recognized by the lesions that appear on the neck, ears, shoulder, and stomach of the pig. They are red and form a diamond shape, but the lesions may range from pink to purple. Once the skin lesions appear, the pig may die in two to four days. Milder forms of the disease do not cause death, but the pig's temperature may spike up to 108 degrees, which can lead to dehydration and harm to internal organs. Penicillin is used to treat swine eryisipelas. If you suspect your pigs have swine erysipelas, you should contact your veterinarian. You will need to discuss ways to eradicate the bacteria from your farm so your pigs are not reinfected.

The bacteria that causes swine erysipelas is always present on pig farms. It is in the soil, is carried by other animals, and is present in the tonsils of pigs. There is little you can do to remove it from your farm. However, there are contributing factors that may make swine erysipelas more likely to appear in your pigs. You may see a surge in swine erysipelas if you have encountered any of the following on your farm:

- If you have wet and dirty pig pens;
- If you give your pigs wet feed, especially if you feed milk byproducts;
- If your pig house constantly has pigs living in it without a chance to air out for a few weeks or months;
- If your waterers carry the bacteria;

- ⑤ If your pigs are otherwise ill or stressed;
- ⑤ If there are sudden changes in temperature with a fast onset of summer weather; or
- ⑤ If you make drastic changes in your pigs' diets.

You may not be able to do much about changes in temperature or summer weather, but you can see to it that your pens are clean, and you can control your pigs' diet, among other things. Try to take care of some of these conditions, and you may keep swine erysipelas from appearing.

Transmissible gastroenteritis (TGE)

Transmissible gastroenteritis (TGE) is one of the most serious diseases found among pigs. It is found mainly in the Midwest among large pig producers. The virus acts very rapidly and causes changes in the intestinal lining, which results in a large loss of fluids. All pigs can be affected, but it is most harmful to piglets younger than 2 weeks old. The fatality rate among piglets of this age may be as much as 100 percent. Symptoms in young piglets include watery diarrhea and rapid dehydration. The disease is highly contagious, and if one litter is infected, then other litters may quickly be infected. Older pigs have a better chance of survival, but they will shed the virus in their manure for months after their recovery. Vaccinations may be effective, but the disease spreads so rapidly that they are often too late to work. The best way to combat TGE is to practice isolating new pigs when they are brought to the farm and to insist that any visitors wear plastic over their shoes to prevent them from tracking germs from other places to your farm.

Parasites

Parasites are an ever-present fact of life on a pig farm. They live in the soil, and they are easily transmitted to your pigs when they root. Parasites are also found in bedding, passed from pig to pig, and exist in manure. Even the cleanest farms with the most well-cared-for pigs will most likely have animals with some parasites. In dealing with parasites, the trick is to keep them to a minimum so they do not overwhelm your pig's immune system and cause it to become ill or weaken it so that it becomes the target of an opportunistic disease. Parasites can be internal or external.

Internal parasites

Pigs can be subject to a number of different worms, such as ascarids, round worms, whip worms, kidney worms, lung worms, muscle worms, nodular worms, pork bladder worms — also known as the human tape worm — red stomach worms, stomach hair worms, thick stomach worms, thorny headed worms, and thread worms. Worms can exist in the lungs, the kidneys, the stomach, the liver, and even on the skin.

Being subject to so many different kinds of worms means that there is no one method of worming — wormers are called **anthelmintics** — that will kill all of the worms or prevent infestation. If your pigs carry a lot of worms, then treating them generally requires a rotational worming program.

There are four basic products used for worming: doramectin, fenbendazole, ivermectin, and levamisole. Although all of these products are very good, one product used alone each time can cause the worms to build up a tolerance, and it will not effectively reduce the worms your pigs carry. In a rotational worming program, a different kind of wormer is used each month until the internal parasites are brought down to a lower number. Once worms are not a serious issue anymore, you can change to worming your pigs once every six months.

In most cases, you will not be able to tell that your pigs have worms unless they carry a heavy infestation of parasites, but you should pay attention to your pigs' physical condition. If your pigs cough, lose muscle tone or body weight, show bad skin, have diarrhea, or show blood in their feces, then you may suspect they have worms. With a very heavy infestation, they may shed worms in their manure. At other times, your only sign may be that they are not gaining weight properly. Any time your pigs seem to be doing poorly without an obvious reason, you should consider whether they may have worms. It is a good idea to have your veterinarian test your herd for worms on a routine basis to see what their status is and if you need to adjust your worming protocols.

Worm medicine can be given to your pigs in various ways, including injections, topically, or in the feed and water. Topical application is usually the least effective of these methods because pigs have a tougher skin and they do not easily absorb things through it. Taking medication orally usually works well as long as the medication is not being wasted in lost feed. You will need to determine the proper dosage to give each pig based on its age and weight.

You can do a great deal to cut down on the spread of worms and other parasites if you will isolate new pigs when you bring them to your farm. Worm new pigs when they arrive, and remove their manure promptly so parasites from the manure do not become established in the soil. You should worm sows before breeding to reduce the parasites they carry and the number they will pass along to their piglets. Once the piglets are weaned, you should also worm them to reduce the parasites they have picked up from their mother. Worm them again two weeks later.

Good pen and pasture management will also allow you to reduce the parasite load. Rotate pens and pastures when possible and allow them to sit empty for several months. This will kill off many of the parasites because they will not find a host. Sunlight will do a great deal to sterilize the soil and make it clean again. When pigs are using pens, clean them out regularly so parasites from the manure do not linger and move into the ground.

External parasites

Pigs are also subject to several external parasites, which include lice and mites, which cause mange, ticks, mosquitoes, and flies. Various insecticides are used to rid pigs of these pests, some of them made specifically for swine. You will need to read labels carefully to understand the precautions and apply them only as directed. Some insecticides have preslaughter intervals, which means you cannot use them within a certain time prior to slaughtering your pigs. Otherwise, the chemicals in the insecticides might remain in the meat when slaughtered. Make sure you have read the label and understand this information.

Treatments for various external parasites can be applied in different ways. There are sprays, dips, pour-ons, and feed additives. Some of the treatments are aimed at treating the pigs' indoor living area. Again, make sure you read the label carefully and apply the product properly, or it will not work effectively.

Insecticides are toxins so you should always make sure you are careful when using these products with your pigs. Keep the products away from children. Store pesticides in their original containers so they can always be properly identified, and wash your hands thoroughly following use. Properly dispose of the cartons after use so animals, including pets, and children cannot get to them.

Tips for Disease Prevention

There are many things that farmers can do to prevent disease and illness on their farms and keep their pigs healthy. In addition to keeping a watchful eye on your herd every day, feeding your pigs well, keeping your pens in good order, and avoiding overcrowding, there are a number of good husbandry practices you can follow that will help prevent disease.

Isolation

Although pigs are very gregarious, social animals, there are times when isolation is in the best interests of the herd. If you have an already-established herd and you are bringing in new pigs, it is in the best interests of your pigs to isolate the newcomers for two to three months when they arrive. They should be kept at least 100 yards away from where your current pigs are kept to ensure the new pigs will not spread any diseases to your herd. It will also give your new pigs time to adjust to life in their new surroundings. If you simply turn them lose into the same pen as your current pigs, you will cause your pigs a great deal of stress, and it will most likely lead to aggression and fighting. Give your new pigs time to settle in while you observe them for any signs of illness.

If you bring sows to your farm for breeding to your boar, they should arrive several weeks prior to breeding so you can monitor them. They will need time to accustom themselves to their surroundings. Some sows may go out of season temporarily when they are moved to a new farm, though they will probably quickly come back in season if the boar is in a nearby pen. Allow the sows and the boar to become used to each other from a distance. If the visiting sows have a few weeks on your farm, they will have time to develop antibodies to any viruses on your farm and will build up immunity before breeding.

Finally, it is a good idea to have an isolation pen or a small pig house where you can keep a sick pig. When one of your pigs does become ill, you will need to remove it from the other pigs immediately so you can treat it and so it will not have a chance to keep infecting the other pigs. Keeping a sick pig in isolation or in quarantine is a good idea from a health viewpoint. Once the sick pig is through using the isolation area, you should be sure to disinfect the area so it will be ready the next time you need to bring a patient to it. A bleach or iodine solution can be used for disinfection, but a peroxygen cleaner, such as Virkon® S, is considered better. Whatever you use, remember that it is vital that you clean the area with a good detergent before you try to disinfect it. It will do no good to disinfect an area that still has debris or fecal matter in the floor. If you are cleaning an open area, you may also wish to consider using a pressure washer, especially if you have been handling pigs with viruses.

Vaccination

Vaccinations can be a good way to protect your pigs against diseases, but it is not recommended that you try to vaccinate your pigs against every possible swine disease. There are too many diseases and vaccines, and most of the diseases will never occur where you live. It is not sensible to stress your pig's immune system with unnecessary vaccines. It is not cost effective to vaccinate your pigs against diseases that do not exist where you live either.

There are several vaccinations that are recommended for pigs at various ages. Getting the right vaccination at the right time can protect your herd and save your litters so it is recommended that you talk to your veterinarian about these basic vaccinations. You should also talk to your veterinarian to find out if there are other vaccinations that are recommended for your specific area. There may be outbreaks of certain diseases and a specific vaccine could be beneficial. Your veterinarian can help you customize an appropriate vaccine schedule for your pigs, depending on your region.

Type of Pig	Diseases to Vaccinate For	Vaccination Schedule
Gilts pre-breeding	Leptospirosis	2 x before breeding
	Parvovirus	
	Erysipelas	
Sows pre-breeding	Leptospirosis	Before breeding and at weaning
	Parvovirus	
	Erysipelas	
Boars	Leptospirosis	2 x per year
	Parvovirus	
	Erysipelas	
Gilts pre-farrowing	E. coli	2 x before farrowing
	Atrophic rhinitis	
Sows pre-farrowing	E. coli	Before farrowing
	Atrophic rhinitis	
Baby pigs	Atrophic rhinitis	1 or 2 x before weaning
Grower pigs (40-100 lbs)	Erysipelas	Time of purchase
(Source: Alabama Cooperative Extension System **www.aces.edu/pubs/docs/A/ANR-0902/**)		

Other methods

Many farmers choose to take a proactive approach to their pigs' health and choose feed and supplements that contain antibiotics, arsenicals, and growth hormones to promote health and growth. Given in this form, these supplements are used to promote growth; arsenicals are also used as an anti-parasitical. This method has been very successful in the pork industry over the last several decades, but it has come under scrutiny in recent years because of the rise of viruses and bacteria that appear to be resistant to antibiotics. It has been suggested that so many antibiotics are used in raising livestock that viruses have become immune to these antibiotics. According to one study by the Union of Concerned Scientists, 70 percent, or 25 million pounds, of antibiotics produced in the United States each year are given to livestock to promote growth.

Some small farmers are trying to use other approaches. **Homeopathy** uses a natural approach, utilizing plants, minerals, or animals, to boost the pig's immune system so it can mount a response to any kind of disease. For instance, a homeopathic approach to diarrhea might assume that if your pig has diarrhea that the body is trying to purge itself of something harmful. Instead of trying to put a complete stop to the diarrhea, a homeopath might try to restore a balance to the body by using particular herbs to help get rid of the harmful substance. There are licensed homeopathic veterinarians in some areas who can work with you and your pigs if you are interested in learning more. You can search for a homeopathic veterinarian on the website of the American Holistic Veterinary Medical Association: **www.holisticvetlist.com/**.

Many people have an interest in herbs. Even if you are not planning to pursue an all-natural approach to raising pigs, there are some good ways to use herbs on your farm. Planting the right herbs in the right place can keep insects away from your pigs; they can make your pig's house smell better; some herbs can have a tonic effect, providing a feeling of rejuvenation; and some herbs are natural wormers. If you would like to know more about herbs and how you can use them on your farm, you can check some of the following websites for more information on herbs:

- Gardener to Farmer (**www.gardenertofarmer.net/herbs/**)
- Alternative Nature Online Herbal
  (**www.altnature.com/herbfarming/considerations.htm**)

⑤ Beneficial Plants (**www.alternativevet.org/Beneficial%20Plants%20WS019-07.pdf**)

Hygiene

Good hygiene is one of the most important ways you can promote good health for your pigs. In addition to taking care of your pens and keeping your farm clean, it is important for you to consider how other people and farms affect your pigs.

Any time you bring new pigs to your farm, you run the risk of bringing disease home with you. If you purchase pigs from sales barns or other places where pigs mingle together from many places, the odds increase that they will exchange diseases. When you bring one of these pigs home with you and introduce it to your herd, you pass along all of the diseases that it has encountered. It is best to buy pigs from fellow farmers who do not pass around pigs very often. This will limit the spread of disease.

Before you purchase a new pig, the seller should provide you with a certificate of health signed by a veterinarian. You may even wish to have your own veterinarian examine the pigs. This is recommended if you purchase purebred pigs or breeding stock. It is worth spending the extra money in these cases to make sure the pigs are healthy and fit for breeding. These extra steps can prevent you from bringing pigs to your farm that may spread disease.

No matter where you obtain a new pig, remember to isolate the pig or pigs for a short time when you bring them home. Pigs do not like to be alone, so it is best if you buy more than one pig at a time. However, it is better if you buy from only one source. If you buy two pigs from Farmer Brown and two from Farmer Jones and another pair from Mrs. Smith, you will increase the likelihood of bringing disease to your farm. Instead, buy all of your pigs from the same source when you buy.

When people come to visit, have them put plastic booties over their shoes. Alternatively, they could step in a pan of bleach to kill germs on their shoes. This will work to kill most viruses. Or you could simply not encourage visitors to your farm, especially if you have a young litter of piglets. You may want to take this route if you have a litter that is less than 2 weeks old, which is when piglets are very vulnerable. Simply tell people they can see the piglets when they are a little older.

After you have raised pigs for a while, you will have your own routine established for pig care. You will have your own ways of keeping your pigs healthy and preventing the spread of disease. These are some basic ideas to get your started. Nothing is worse than losing a herd of pigs after you have invested time, money, and so much of yourself so be very careful with them.

When to Call in the Vet

There is currently a shortage of large animal veterinarians in the United States. The American Veterinary Medical Association (AVMA) anticipates that this situation may continue for some time to come. This means that, depending on where you live, you may or may not have easy access to a quality large animal vet. If you are just starting out raising pigs, it is a good idea to try to find a large animal veterinarian before you purchase animals. To find a large animal veterinarian in your area, you can look in your Yellow Pages, check online, or ask other farmers to recommend one to you. You can check the website for the American Veterinary Medical Association (**www.avma.org/statevma**) for veterinarians in your area, but not all veterinarians belong to that organization. Find one, talk to him or her, and invite him or her to visit your farm. If they are experienced, they may be able to offer you some good advice before you start raising your pigs. It is also a good idea to get to know your vet before you have an emergency.

Once you bring your pigs home, you will probably need to have your veterinarian come out to perform a few vaccinations, whether they are grower pigs or pigs you are preparing for breeding. Most veterinarians like to visit a herd when it is healthy so they can size it up. This gives them something to use as a baseline in case they need to visit when some of your animals are sick. They can also talk to you about preventive care and get to know you better. Having a good relationship with your veterinarian is important for both you and your vet.

If you have a sick animal, it is important that you do not wait too long to call the vet. Although there is no need to call the vet every time your pig sneezes, it is important for you to know your animals well enough to know when something is really wrong. If your pig has a fever or diarrhea, it is probably a good idea for you to go ahead and call the vet, especially when you are not experienced with raising pigs. Many swine diseases have symptoms that may be very serious. Some of them can cause young pigs, especially, to become dehydrated very quickly. It is best to take no chances and call the vet. If the vet says there is nothing to be worried about, you will feel better. And, if it is a serious illness, the vet will know how to treat it. As you become more experienced in raising pigs, you may be able to distinguish a serious swine disease from one that is not so serious, but when you are just beginning, it is best to err on the side of caution instead of risking your herd.

If you have questions about when you should call your veterinarian, the best thing to do is to ask him or her for some guidance. A sow can deliver a litter by herself, and there is no need for a veterinarian to be present. Animal husbandry practices on the farm, such as cutting wolf teeth or castrating young males, are things you can do yourself, though you may want another farmer to show you how. Potential diseases, however, can be serious. They can often affect your entire herd so in this situation, it is best to call the vet.

Summary

In this chapter, you have learned about pig diseases, their treatments, and possible ways to prevent them. Some of the diseases are very serious, and there may not be any treatments. In other cases there are effective vaccines. You should talk to your veterinarian or a county extension services agent and find out about the vaccinations recommended for pigs in your area. This chapter also discussed hygiene and some of the things you can do on your farm to keep your pigs healthy by keeping things clean. Good animal husbandry is one of the best means of preventing diseases.

In the next chapter, you will learn more about herd management and ways you can keep your farm better organized. Good organization is very important for anyone who plans to keep more than one or two pigs because good organization helps you provide the best possible care for your pigs.

CHAPTER 8:

Herd Management

Now that you have considered some of the basics of pig care, such as housing, feeding, and health, it is time to look at herd management. Successfully managing your herd will entail keeping track of daily tasks, such as how much you feed, how much weight the pigs gain, when you worm them, what vaccinations they receive, and the other things that go into raising pigs. It will also include looking at the big picture at the end of the year and learning which things worked well and which things did not.

Managing your pigs falls into two areas: the day to day operations on your farm, such as feeding, cleaning out pens, moving pigs from place to place; and managing things on a larger scale. This can include keeping track of which sows produce large litters, which boar breeds well, which feed works well for your herd, and other things that can affect the long-term interests of your operation. Both parts of your management are important and will determine how well you do with your pigs.

Daily herd management tasks mean ensuring your pigs are not unduly stressed and ensuring they eat well and maintain good health. For this to happen, your pigs need to be maintained in compatible groups. Whether you raise grower pigs for market or keep several sows for breeding, it is important they get along reasonably well. If you have an overly dominant or aggressive pig, it is likely to cause an undue amount of stress to the other pigs, which will affect their health and weight. You can minimize these problems by separating pigs.

Separation

There are many occasions when separating pigs is a good idea. When you bring new pigs to your farm, it is recommended that you isolate them for 30 to 60 days at a distance of at least 300 feet from your main herds. This will prevent any diseases they may harbor from spreading to your pigs. You can minimize their sense of isolation by purchasing more than one pig at the same time.

If you bring in gilts or sows for breeding, it is a good idea to separate them from your herd as well. As you approach the time you intend to breed them, you can move them closer to the boar's pen so they can become acquainted. If you introduce new gilts to your herd then, following the quarantine period, you will need to pen them near the main herd for a few days so they can start to know each other. These introductions do not always go smoothly so you should watch carefully once the new gilts have been introduced.

Boars must be separated from the main herd and kept by themselves. Boars can be very aggressive, and it is in the best interest of the herd to live without the boar on a daily basis. Boars should not be kept together or with barrows.

If you raise grower pigs for market, then you can raise young gilts and barrows together. They are often from the same litter, and females and castrated males get along well. Other farmers prefer to separate gilts and barrows when they grow them out because they think that gilts grow better when kept in a same-sex pen. Farmers can also adjust the feed ration to make it more favorable for gilts when they are separated. Gilts can do better on a feed ration that has more amino acids, especially lysine, than barrows.

Another time when you will want to separate your pigs is when you have young litters of different ages. A mere two- or three-week difference in age can make a tremendous difference in status and behavior. If you place 5-week-old piglets in with 7-week-old piglets, the 7-week-old piglets will make life miserable for the younger piglets Just two weeks' difference in size will be enough for the older pigs to dominate the younger ones physically. They may pick on them and be aggressive toward them, which can stunt the growth of the younger pigs. They will continue to dominate them as they grow older. It is best to separate them until they are more mature and there is less difference in their size.

Size differences and the aggression and stress they can cause in young pigs have led many pig farmers to use the **all-in and all-out method** of pig production. With this method, all of the sows due to give birth within a few days of each other are brought into the farrowing area at the same time — "all in." Their piglets will all be born around the same time and will all be about the same size. It will be all right for them to play together, feed together, and live together later. When they are ready to wean they can all go out at the same time — "all out." By that time, it should be time to bring in the next sows ready to farrow. With this method you always keep pigs of the same age and size together and avoid many herd problems. If you have two sows with litters of the same age, it is fine to let them reside together in the same pen. Sows will normally take care of each other's piglets and defend them if necessary.

Daily Operations on the Pig Farm

Pigs are very easy to care for, which means your daily routine with your pigs does not need to be extremely time consuming if you only have a few pigs. Your daily routine will vary somewhat, depending on the age of your pigs and your plans for them. For example, if you get your pigs ready to show at a fair, you will probably need to spend some time grooming them, and you may want to hand walk them for exercise. You will probably hand feed them — in a tub — instead of feeding them using a self-feeder or raising them on pasture in order to carefully monitor how much and what your pig eats. On the other hand, if you raise your pigs for slaughter, your routine will be very different. If you raise your pigs on pasture, your daily operations might consist of nothing more than going out to check on the pigs, making sure their water source works, and planning to supplement their pasture grazing with some grain later in the day.

If you raise your pigs in a large pen, you will probably check their self-feeder first thing in the morning to make sure it works and the pigs have plenty of feed. You will need to check their water supply. And, you will need to clean the manure out of the pen, which may take about five to ten minutes.

You can also spend some time petting and scratching your pigs. Time spent socializing with your pigs is always time well spent. Your pigs enjoy this time, which lowers their stress levels. This helps them feel better, eat better, and grow better. Spending time with your pigs also makes it easier to work with them.

Daily operations could also include noting any changes in the feed you give your pigs, making note of any worm medication you give them, or noting their physical condition if they might need worming. You should keep a record of any changes in your pigs' health or condition, for better or worse.

If you receive any shipments of feed or you purchase any feed, you will need to ensure it is stored properly and make a record of receiving the feed and the price you paid for the food. Naturally, if a sow farrows a litter or you have a breeding, you will need to attend to those matters. Such incidents provide more excitement than usual daily farm operations. Most farms are set up so the pigs can either graze or feed themselves at the self-feeder so there is no need to plan for an evening feed. Pigs are not normally bedded down at night the way some stabled animals are so there is no need to put them up. As mentioned, it is very easy to care for pigs.

Tracking

Keeping track of your animals will become a very important part of herd management. Although there is not currently a national tracking database for farm animals, there has been much discussion of developing such a database for tracking disease transmission and for biosecurity purposes. If animals are tagged with a national database ID number, the government would always be able to identify where they originated, where they were sold, and what their final disposition was. By extension, this would also allow the government to monitor the nation's food supply more closely. However, the system has many critics because it would mean giving up a great deal of personal information on the part of farmers and could be very intrusive. It is possible that such a database may be instituted in the future, or it may not be.

Being able to track your animals in some accepted way allows you to report on them to purebred registering bodies and to report about them to your county or state in case of disease outbreaks and allows them to be identified when they are sold. Being able to easily identify your pigs is also helpful when it comes to keeping records about your herd.

A word about the importance of accounting and records

It you have one or two pigs, it may seem unnecessary to worry about accounting or keeping records. But, if you plan to continue to raise pigs, it is a good idea to get in the habit of keeping track of how the money goes in and out where your pigs are concerned. It is also important to keep good records about your pigs whether you have one pig or 1,000.

In terms of recordkeeping, you should make note of each change of feed, what you feed, what supplements you use, and how your pigs do on the feed. It is a good idea to keep a card or other record for each pig. You should keep track of the vaccinations the pig receives; illnesses; when the pig was wormed; and the date obtained or whether it was farrowed at your farm. And, you can note if you sell the pig or have it slaughtered. If the pig is slaughtered, you can give an account of the final weight and any notes about the meat and fat. All of these records can provide you with important information so you can improve your pig operation in the future. If you breed your pigs, you should keep records about sow litter size, losses, how fast the piglets grow, and how the piglets develop later. By following this litter information, you can determine which sows are the best mothers and which breedings you need to repeat. You can also determine which sows you may need to cull.

Earmarks and tagging

Earmarks and tagging provide good ways to individually identify each pig. **Earmarks** or ear notches are a traditional method that has been used for a long time. The notches are made according to a system and chart with the litter marking on one ear and the individual pig marking on the other ear. By knowing the chart system, anyone familiar with pigs can identify the pig. Each different notch indicates a number, depending up on where it is placed on the ear. Notches for litter numbers are placed on the right ear and notches for individual pig numbers are placed on the left ear. Ear notching is done when pigs are just a day or two old when there is the least chance of causing them pain.

Tagging is done by putting a small tag with an identification number through the pig's ear. It is a simple procedure and is usually done when the pig is very young. The ID number allows you to identify the pig by checking the number. Tags may be colored to allow pigs to be more easily identified if they come from different sellers or if they are different ages, for instance. You may also use photos or tattoos to help you identify your pigs for recordkeeping purposes. A photo is a good way to keep track of how a pig develops from one growth period to the next. It can be kept with your other records for that pig. Pictures can be surprisingly helpful even though pigs look similar.

Software

There are a several swine management software programs available to help pig raisers manage their herds. They include PORCITEC from AGRITEC Software (**www.agritecsoft.com/en/**), Swine Management Services, LLC (**www.swinems.com/**), and PigCHAMP (**www.pigchamp.com/**). Most of them are designed for larger producers, but some are aimed at smaller farmers, and you may wish to look at them to see if you would like to use one of these programs. Swine management software can do analysis to assist you in buying and mixing feed more economically for your operation or can help you analyze your herd's production. The software comes in different formats, such as farrow-to-finish — birth to slaughter — or grow-to-finish — purchase as a shoat to slaughter — depending on the farmer's goals. If your herd is large enough, it could be a good investment for you to consider swine management software for your recordkeeping and analysis. You might need to have 30 or more pigs before this kind of software would be cost effective.

What Your Peers May Expect of You

In discussing herd management and how you maintain your farm, it is also a good time to bring up the subject of your neighbors and your peers. Although you may be in charge on your farm, it is impossible to raise pigs or any livestock without drawing scrutiny. Pigs do prefer to be clean, but there will always be a certain amount of odor associated with raising them. Unless you have a very large property, your neighbors will be well aware that you raise pigs. As someone who raises pigs, it is important for your own sake, and as a representative of all pig farmers, that you take care to manage your property and your pig waste well. Make sure you pick up manure in your pig pens and manage pastures where your pigs may graze. Care for compost heaps, be mindful of water runoff, and do not pollute creeks and streams. Pay attention to the cleanliness of your farm and your animals. People do judge you by the impression you and your farm make. You do not have to have all new equipment, but you can keep your equipment clean and in good shape.

You are also a representative of all farmers in the way that you treat your animals. Most pig raisers and farmers are devoted to their animals and provide excellent care, but it only takes one neglectful person to give all pig raisers and farmers a bad reputation in your area.

Try to follow good breeding and pig raising practices at all times. Your fellow pig raisers will base their own opinion of you on your practices. In most places, the pig farming community is relatively small and word will get around about how you raise your pigs and care for them. You will likely have many casual encounters with others in the pig farming community so try to represent yourself, your farm, and your pigs well.

Summary

This chapter examined herd management techniques. It is very important for you to be able to separate different groups of pigs at different times, whether it is a sow and piglets, new gilts and sows, a sick pig, a boar, or piglets of different ages. You should keep a good recordkeeping system, whether you use computer software or personal note cards. You also need to make sure pigs can be identified in some way, especially if you have purebred pigs. You should identify them with ear notching if you plan to register them with a purebred registry. Finally, you considered what your neighbors and peers may expect of you.

In the next chapter, you will learn about showing pigs. Whether you are interested in pigs as a 4-H Club or FFA project or raising pigs as show stock, there is a very active show world devoted to showing pigs. Many people enjoy going to hog shows, showing their pigs, or simply viewing the latest hog trends.

CHAPTER 9:

Going to the Show

Up to this point, you considered pigs mostly from the viewpoint of raising them for food, whether for your own table or to be sold to others for pork. However, there are many people who enjoy raising pigs for another purpose. There is a thriving show scene for pigs at county and state fairs and at market hog shows. Enormous prestige goes along with winning classes at these shows or with raising a hog that earns the title of grand champion of a fair.

In this chapter, you will consider why people show pigs and what benefits they may obtain from it. You will look at how to choose pigs for showing and how to raise them so you can hope to do well in shows. You will look at some of the things you can do to present your pig, and yourself, well when you are in the show ring. And, you will consider what happens after the show and what you may learn from the experience, especially if you are a young exhibitor.

Why Show Pigs?

If you are new to raising pigs you may wonder why people show pigs. What is the purpose? After all, there are computer programs to track production values for sows and their litters, the production records of boars, rate of gain, and other important considerations for pig farmers. Why is it important to measure pigs against each other in a show ring?

Long-time pig farmers will tell you that it is important for people who work with pigs to have a good eye for a pig, or for any kind of livestock. People who work with animals should be able to judge their value by looking at them and sizing them up properly. The only way to develop that eye is by watching countless pigs, hearing the opinions of experts, seeing pigs evaluated against each other, and learning what to look for in a good animal. There are no shortcuts to this kind of education in livestock knowledge; it is only gained by hours and years spent studying animals. This is the kind of knowledge gained at shows where pigs are exhibited and opinions about them are exchanged. The same is true for shows where cattle, horses, goats, lambs, and even dogs and cats are exhibited. People knowledgeable about the animals come together to share information and pass it on to others, especially to newcomers. There is much to be learned by anyone interested in learning. In many cases, someone who has a good eye for one kind of animal will also be able to select good animals of other species. A good hog person will be able to recognize good cattle, for example.

Of course, you could show other animals besides pigs at a fair or livestock show, but pigs are a very popular choice. It is less expensive, and there is less equipment involved with showing pigs than with other animals. It is less costly to purchase a young pig than a calf, for example, and though a young lamb is not expensive, there is much more equipment needed to show a lamb than a pig.

FFA and 4-H Club

The National FFA Organization and 4-H Club both promote youth involvement with agricultural projects, which include showing pigs at livestock events, such as fairs. FFA was founded in 1928. At one time it was known as Future Farmers of America, but today, it is simply FFA. Today, FFA is a career and technical organization for students focusing on agriculture. It is open to students in middle and high school classes. In addition to production agriculture, FFA also provides educational opportunities for teens interested in food, natural resources, and science and technology. FFA has some 520,000 members. There are more than 7,000 chapters in the United States, Puerto Rico, and the Virgin Islands. Membership is for people starting at age 12 but not beyond age 21. Involvement at the local chapter level is strongly emphasized and participation is through schools. FFA is part of the educational system in school classrooms and includes classroom instruction, as well as hands-on agricultural experience. If your school does not have an FFA program, you can contact the FFA agricultural staff in your state to find out how to start a school program. You can also join an FFA chapter at another local school.

The 4-H Club is a youth organization for people ages 5 to 19, though there are also collegiate 4-H clubs for alumni who wish to remain active with the organization. Various 4-H clubs began to form around 1902 as after-school agricultural clubs. By 1914, they were a national club. Congress included 4-H Club when they created the Cooperative Extension Service of the USDA in 1914. The 4-H is administered by the USDA. It has some 6.5

million members in the United States, and there are more than 90,000 4-H clubs. There are also 4-H clubs and organizations in other countries now. There is a fee to belong to a 4-H Club, ranging between about $20 and $40 depending on the local chapter. Sometimes the club may have fundraisers to raise club funds. Members pay for their own project costs and fees. The goals of 4-H are to develop leadership and life skills in youth. In order to do this, 4-H uses learning programs that emphasize experience and tries to develop good citizenship skills. Along with agriculture, the organization also promotes healthy living, and encourages kids to become involved in science and technology. The motto of 4-H is "to make the best better," and their slogan is "learn by doing."

Both FFA and 4-H feature livestock projects in which kids can raise a market hog from the time of weaning until it is market hog weight — about 220 to 260 pounds. Kids can also take part in a project where they care for a gilt-and-litter or sow-and-litter of pigs from the time of mating until the pigs are ready to sell when they are a few weeks old or until some of the pigs are market weight. Both kinds of projects are excellent learning experiences for youths, though raising a weanling to market weight is more popular at the present time.

Once the pig reaches market hog weight, or the gilt-and-litter or sow-and-litter reaches the time when the feeder pigs would be sold, the animals are exhibited. Of course, these are not ordinary pigs. They have been specially chosen as show prospects, hand fed — their feed carefully mixed and perhaps fed to them singly in tubs instead of in an automatic self-feeder — and their weight and condition carefully watched to make sure they have developed to their full potential.

CASE STUDY

Sean "Patrick" Weick, II
1st time 4-H member
Student
Mt. Hermon, LA

CLASSIFIED CASE STUDIES
directly from the experts

Patrick Weick is 9 years old and on his third set of pigs he is raising as a 4-H Club project. "My first set was a Landrace/Yorkshire cross. I raised them out to 150 pounds live weight to see if I wanted to do this as a 4-H project. We sent Gilbert and Spot [the names of his pigs] off to the slaughterhouse when they reached 150 pounds and got two hams, pork chops, and the rest in smoked sausage. In June 2010, my parents bought me two three-quarter red wattle and one-quarter Hampshire gilts — Ruby and Garnet — for me to raise and to show at the Washington Parish Free Fair in 2010. Then, Garnet started biting, and she was sent to the slaughterhouse and was dressed out at 116 pounds. We got one ham, pork chops, and the rest in sausage." He said he has plans to breed his boar to Ruby.

According to Weick, the benefits of raising pigs are that you know what goes into them and "when you get your meat back [from the slaughterhouse], you will never go back to store-bought meat, and that's a promise."

Weick says if you are interested in getting started in pigs, "You need to purchase two, or they will not grow as well for you [because they are lonely]. Don't follow in my footsteps and buy the biggest one and the runt. You will set yourself up for a big fall if you are keeping track of their weight gain."

Weick, who lives in Louisiana, is raising his pigs outdoors. "The [pigs] are outside in three pens that are 16 feet by 16 feet with a shelter and a roof over half of the pen. Right now due to the temperature, the Herefords are together to help keep each other warm and would be moved to their own pen at the start of February. They are on ground, and when [the pigs] have babies of their own, I will not have to worry about an iron shot to the piglets because they will have access to dirt. Sometimes I wished we made the pens smaller, and sometimes the pens are the right size when you have two 125-pound pigs in the same place."

As for regulations and veterinary care, Weick has this advice: "Get live-stock insurance on them just in case they get out and get hit by a car. That was the first thing my parents did before Gilbert and Spot came onto the place. We do have a vet that we can use if we need one, but I'm trying to raise my pigs as all natural. I have a sheet in my 4-H book [with recommended shots], but after we thought Summergirl had mange, the ivomec shots racked my nerves. We gave 1cc per 75 pounds every two weeks for six weeks. It cleared up her hair problem."

Weick has already raised several different kinds of pigs. "I have had cross (Landrace/Yorkshire), hybrids (three-quarter red wattle and one-quarter Hampshire), and heritage (Herefords). If you want a fast growing pig, get something that has Landrace or Yorkshire in it. Ruby, the three-quarter red wattle and one-quarter Hampshire, grew slower but when we took her sister off to the slaughterhouse, she had leaner meat than the Landrace/Yorkshire cross. Some say the Herefords aren't that great when slaughtered, and I hope that we never find that out."

Weick says that the most important thing to know about raising pigs is that you should not get just one. They are very social animals. "Don't just get one pig, or they will be depressed, and when you have a depressed pig, you will soon enough have a sick pig on your hands. Spend time with them each day. They love treats and for their ears to be scratched, and they are smarter than a dog."

Weick has some funny stories about his pig Ruby. "I don't know who taught who about how to walk a pig, Ruby teaching me or me teaching Ruby. But, after the first week of walking her out of the pen, she knew which pecan tree had good pecans on the ground and which tree to pass by. Ruby never liked my show heifer Summer who is a polled Hereford. When walking, she always gives her her space. But, when Ruby had enough walking out of her pen, she would go back to her pen and shut the gate in my face and stand in front of her bowl for feed. Even though the next time I will show Ruby will be in October 2011 at the Parish fair, we still go for our 45-minute walk around the place. She is my best herd pig when it comes to moving the cattle from one field to another. Ruby loves to show her teeth, but if she grunts while showing her teeth, back off or she will snap at you. Ruby was easy to teach dog tricks, and my mom got mad at me and asked what I would do if she sat in the middle of the show ring and started to bark. I told her I would pat Ruby on the head and whisper in her ear that the marshmallow [treat] would have to come after we got judged."

Weick is very wise for a 9 year old. When asked if he has additional tips about raising pigs, he said, "Don't stress out over the little things with your pigs because when you are stressed so are they, and they don't recover as quickly as you do. Respect them while they are growing and when it's time to slaughter them; their meat won't be tough. Give thanks for that animal every time you eat something from them, and remember the good times you had with them, and look forward with the next set you have. The quality of life is way better than the quantity of life."

Personal Development

Both FFA and 4-H Club do a great deal to promote personal responsibility among their young members. Being involved in livestock projects and being responsible for the care of animals is a good way to teach youths about animal care, commitment, and many important life lessons. The pigs will depend upon their young caretakers, which may be a new experience for some kids. Kids learn to be responsible for feeding, grooming, and conditioning their pigs, and they may also learn important lessons about compassion. They will probably learn some lessons about winning and losing and how to deal with their emotions. They will learn about competition, and they should develop more self-confidence.

People involved with FFA or 4-H, or even those who simply attend various fairs and get to know other people who show pigs, will make many friends their own age with similar interests. If they continue to show different project animals year after year, they will learn not just about pigs but also about each other and about themselves.

Profit

Unlike pigs bred simply for pork, the profit in show pigs lies in the genetics and in breeding to win. There are many pig breeders who breed primarily for show ring success. A **proven sow** — one that has already produced champion offspring — may be worth several thousand dollars. Boars who are successful sires and who have produced pigs that have been successful in the show ring may command high prices at stud. They are often available to other breeders by means of artificial insemination because they are simply too valuable to allow to leave their farms. Plus, they may be in demand as a sire all over the country. Boars such as these will often have names that are recognizable to people who regularly attend shows where pigs are shown or to people who are familiar with pig genetics.

Farms that specialize in show pigs often take orders for barrows and gilts for the upcoming year and work with families seeking project pigs for their children. They know the kind of pigs that will catch the judge's eye and win in the show ring.

These project pigs, or young feeder pigs, are sold at weaning time. They will cost more than a typical feeder pig you might find on a farm for pork production because of their genetics and the work that has gone into raising and showing their sire and dam. These young feeder pigs have a pedigree, and they already have a reputation. They have also been carefully bred based on their parents' appearance and ability to win. They are expected to grow up and do well in the show ring for the person who purchases them, provided they are raised and fed correctly.

If you, in turn, wish to make a profit from your pigs after you finish showing them, they are often worth more if you sell them or use them as breeding animals than for their pork. Depending on whether they are crossbred or purebred animals — some first generation crossbred pigs can do very well in the show ring if it is an attractive and popular cross — you may be able to use a good gilt as the beginning of a breeding program or fit her into your current herd for use with a very good boar. Or, you could sell her to someone else who can use her for breeding. If you show young feeder pigs in a gilt-and-litter or sow-and-litter class, you may wish to grow them out to show later in the year at other shows as market hogs. Or, you could sell them for a good price to someone else who would like to show them as market hogs. There are often many options for good animals that do well at shows.

The Show Pig

Popular show pigs today can be found among the more popular breeds of pigs, such as the Yorkshire, Chester, Hampshire, Duroc, Landrace, and Berkshire, with some spots here and there. Hampshire crosses are also very popular, such as the Hampshire x Duroc and the Hampshire x Yorkshire. These crossbred pigs often have a very good chance of winning if they have the right look and have been raised and cared for well.

Before you go looking for a project pig, you will need to decide which class you are aiming for. Are you looking for a feeder pig to raise for the market hog class? Or, are you looking for a gilt or sow to breed or that has been bred so you can show her and her litter in the gilt-and-litter or sow-and-litter class? Obviously, the latter choice will cost more because you would be buying an adult pig or one that had been bred. There is also much more work involved in caring for a gilt or sow with a litter of piglets. This is often a project for an older teen who has moved beyond the market hog class.

If you choose to show in the popular market hog class, you will need to find out when your shows are. You can check in your local feed store for fliers about upcoming events, hog shows, fairs, and livestock shows. There is a fair circuit and fairs in different counties and states are normally held from August through October. There may also be some market hog shows in your area. You should decide which shows you plan to attend so you can decide when you would like to purchase your pigs. For example, pigs raised to be shown at the August to October fairs are normally born between December and March or April so they will have several months to grow and be at optimum weight for the shows.

Breeders who breed show pigs will have barrows and gilts available at the right times for people interested in purchasing project pigs. If you plan to purchase a pig from a show pig breeder, you should contact them early to discuss the litters they are planning and the piglets that may be available. You may want to contact a breeder who has a proven record of producing winning pigs in your area. Judging can vary in different parts of the country, and you will want to have the kind of pig that can win where you live. On the other hand, if you have a definite idea of the kind of pig you like, you can contact a breeder who produces that kind of pig. Good pigs will often win anywhere. It is best, however, to avoid extremes of type, such as a pig with an extremely long body or overly enlarged hams. A pig should not look exaggerated or like a caricature of its breed.

This brings up the issue of type, or breed type. **Type** is a concept that is often hard to define but easier to point to when you see it in a nice pig or any other animal. Type is the essence of a breed. It is a combination of those characteristics that people associate with a breed, such as the rich red color of the Duroc, the drooping ears, the shape of the body, and the personality. Type is what distinguishes one breed from another so you do not mistake a Duroc for a Tamworth or a Chester for a Landrace.

A show pig should have plenty of breed type. It should be immediately recognizable as whatever breed it is. It should be a good physical representative of its breed, as well as healthy and in good condition. As with developing an eye for good animals, the longer you spend looking at livestock, the easier it will become for you to see good type when you find it. Some animals may have all of their parts in the right places, and they still may not have good type. They may lack some important intangible quality or qualities. There are many subtleties of type, and it does take time to see some of them.

Choosing Your Show Pig

When it comes to choosing your show pig, it is a good idea to do some homework before you pay $10 to $20 per pound for a young feeder pig that weighs 40 pounds. Young show pigs cost quite a bit more than ordinary feeder pigs so you need to be clear on what you are looking for in a pig.

You should start by studying breed standards for the breeds that interest you. All of the major pig associations have standards that describe how ideal pigs of their breeds should look. You can also find recent pictures on the Internet of pigs that have won at shows. There are also some good magazines that report on hog shows, such as *The Purple Circle* (**www.purplecircle.com/**) and *Seedstock EDGE*, the official publication of the National Swine Registry (**www.nationalswine.com/Home_pages/about_NSR.html**). By looking at winning pigs and understanding breed standards, you will have a better idea of what a good pig should look like.

When choosing a pig, it is also helpful to visit local hog shows in your area and see what kind of pigs your local judges prefer. Watch them judge classes, and see what kind of pigs they select. Listen to their explanations if they offer details about why they chose certain pigs as their winners. See if their selections are consistent for certain traits. Local judges are normally county extension service people, experienced hog breeders like those who may breed purebred pigs, people from the meatpacking industry, and university agricultural personnel so they represent a wide spectrum of hog experience. They may have different opinions about pigs, and a winner one day may not win the next time out. You should keep that in mind when you show your pigs, especially if you do not take home a blue ribbon. You are always asking a judge for an opinion when you show your pigs, and some opinions are worth more than others. If you do not win one day, there will always be other shows.

After you have a good idea of what a good pig should look like and what kind of pigs are winning in your area, you should be ready to go looking for a show prospect. There are pros and cons to selecting both gilts and barrows. Gilts grow more slowly than barrows, but at the same time, they also tend to be leaner, which is an advantage. Being leaner allows you to feed them feed that is higher in protein, which can put a great finish on them for the show ring.

Barrows will grow faster than gilts, but you will need to be careful that they do not become overly large before the show. Pigs in the market hog classes are shown at around 220 pounds to be competitive and should be leaner rather than fatter. You will have to carefully monitor their feed, and you may have to cut their protein percentage by 1 or 2 percentage points if they appear to gain weight too fast.

Whether selecting a gilt or a barrow, choose the largest and best-developed pigs available. This is a good indication of how the pigs will develop over the next few months as they grow out. Make sure any barrows are not coarse looking and that they do not appear flabby or fat. They should look trim, especially around the head and underline. You should be able to count several nipples on the barrow in front of its sheath. This is a good indication that the barrow will have a desirable long body as an adult.

Both purebred pigs and crossbreeds are doing well in the show ring right now. Choose a pig that you like and that you think you can win with. If you are proud of your pig and have confidence in it, it is often easier to put in the hard work necessary for a project pig and to present the pig well in the show ring. If you get a pig you do not particularly like, chances are that caring for the animal will seem more like a chore, and you will not enjoy presenting the pig as much.

If you choose a crossbreed, it is a good idea to choose one that is particularly eye-catching. Choose one that is black or red. Crossbreeds that are black or red feature good muscling that comes from their colored parents. Their color and muscling will catch the judge's eye. You may also wish to consider the show conditions when choosing your pig. Black and red pigs may show up better under artificial lighting if the show is indoors. White pigs may look better in daylight if the show is outdoors. You can find out all about the upcoming show conditions by consulting the premium list for the shows you intend to enter or by simply talking to people who have entered those shows previously. You can obtain premium lists by contacting the superintendent or show secretary who will be in charge of holding the show. There may be a club in charge of holding the show, but they normally designate one person or agency to field questions and take entries.

Finally, you may wish to raise two or even three pigs as part of your project, in case one of the pigs does not live up to expectations. It is more expensive to purchase extra pigs, but it offers a good fallback for you in case of any health problems or other unforeseen troubles. It is often a good idea to bring a backup pig with you when you go to the show and weigh in with your pig.

If you follow these suggestions, you should be able to find a good show pig. You should do the same things you would do when buying any feeder pigs:

- Make sure the pigs have a health certificate.
- Make sure they have been wormed.
- Ensure the farm where they have been raised meets your expectations for hygiene.
- Discuss pedigree and how the pigs have been raised with the breeder.
- Keep your pigs vaccinated for diseases found in your area as your show pigs will be traveling and coming into contact with pigs from various places.

Feeding for the Show

Feeding your young show pig for shows is different from feeding ordinary feeder pigs. Even more so than with ordinary feeder pigs, you will be feeding your show pigs for optimum weight. It is not unusual to feed your show pigs a premium feed that is 15 to 16 percent crude protein throughout the period leading up to the shows. Some people even choose to feed a higher crude protein percentage and feed 17 to 18 percent crude protein feed. There are commercial feeds for show pigs that contain this amount of crude protein, or even higher. Lindner Feed is one such company (**http://showfeeds.lindnerfeed.com/pid1pigfeeds.html**). These kinds of feed do cost more than ordinary commercial feed. Bags of this show feed that contain higher crude protein levels and other ingredients aimed at getting faster growth may cost between $20 and $30 instead of $10 to $15 for ordinary swine feed. They may contain higher — or lower — fat content and specific minerals aimed at certain growth periods in a show pig's development when you need to increase or decrease their gain.

There are certain challenges involved in feeding show pigs. Because of the fact that you will most likely try to encourage your pig to gain weight over the hottest summer months, it is very possible your pig may cut back on how much it eats at times because of heat stress. This is not unusual for pigs. Many pigs eat less when it is very hot — more than 80 degrees, especially when it is very humid. However, when you feed a pig in the hopes of winning at shows, it can be a disaster if your pig loses interest in its food or, even worse, starts to lose weight.

There are some ways to get around this problem. Many people raising pigs for shows will change their feed from grain form to pellets. Pellets seem to be more appetizing to pigs and will tempt even pigs who are not much interested in food. They are also very easy for your pig to digest, and they are densely packed with nutrition. So, even if your pig eats less of them, it will get plenty of good nutrition. Increasing the protein content during hot weather is another option to make sure your pigs get plenty of protein for growth.

You can also try adding extra fat to your pig's diet. This is a common approach many people raising pigs for shows use. Fat is concentrated calories, and there are several commercial show pig feeds that contain higher fat content just for this reason. It is not unusual to add 3 to 5 percent more fat to the pig's diet or to dress their ordinary meals with a liquid cooking oil. If you plan to add fat to your pig's diet, you will need to be careful to make the change slowly because adding fat can cause gastrointestinal problems, especially with pigs being so carefully fed.

If your pig shows a lack of interest in its food, you can also add some pig starter mix to its ordinary diet. This is the feed that is ordinarily fed to piglets when they are visiting the creep and still nursing from their dam. It is very nutrient rich and contains milk products. It is also expensive so this is not something you will want to continue for long periods. However, if it gets your pig eating and through a period of hot weather, it will have served its purpose.

On the other hand, if your pig eats too much or gains too much weight too fast, it is a good idea to cut back on its rations a bit. Try reducing your pig's crude protein intake by a percent or two if it is eating a high percentage. If it eats from a self-feeder, you may need to remove it and start hand-feeding the pig or feeding it by using a tub so you can control its portion size. The pig may be eating all of its feed and another pig's, too.

Remember to make any changes to your pigs' feed very gradually. You are feeding for optimal growth, appearance, and good health. Any sudden changes could stress your pigs and cause a setback.

Fitting Your Show Pig

In its broadest possible meaning, fitting your show pig refers to all the things you do to prepare your pig for its appearance in the show ring. You are really fitting your pig for exhibition from the time it is destined to become a show pig. If you maintain this attitude toward your pig, it can help you keep your focus on the importance of everything you do to raise your pig for showing. But, in a more precise sense, "fitting" your pig refers to the specific things that have to be done to prepare your pig to go into the ring, such as training your pig to respond to the show stick and bathing and grooming your pig.

Fitting your pig also means that your pig should be in good weight and condition for the show ring. By the time your pig weighs about 175 pounds, which is about four weeks before the time your pig makes a show weight of 220 pounds, you will need to start considering your pig's "finish" or finished appearance. Leaner pigs are preferred over pigs that carry more fat so you will want to make sure your pig has plenty of muscle mass without looking chubby. You should continue to feed for your pig's optimum weight and size while making sure it is in good condition. In order to do this, you may need to start increasing your pig's exercise.

You can increase your pig's exercise by hand walking it. Your pig will need to know how to walk calmly next to you in the show ring. You can guide your pig by using a cane or show stick and gently tapping it behind the shoulder. Your cane or show stick should be a plain but attractive item. You can purchase a 37-inch cane for around $16 from a site like ScottDale Supply (**www.scottdalesupplyonline.com/pig_show_supplies_i.htm**). They also have shorter canes if you are not as tall, and they have pig sticks if you prefer those. Of course, it will take practice at first for your pig to learn what you want it to do, but pigs are very intelligent, and your pig should learn quickly with some training. It is a good idea to practice in an enclosed area, such as a pen or a fenced, grassy lot. If you practice walking with your pig a little each day, the two of you should begin to move as a team, and your pig should get the extra exercise it needs to look good for the show ring.

You can also have someone watch you walking with your pig so you know how your pig looks best. You can simulate the show ring in your pen and move your pig in a large circle. Does your pig look best from behind? From the side? You can practice showing off your pig's good points and learn not to emphasize its not so good points.

You will need to consider variables like the weather when you are fitting your pig for the show. Do not try to exercise your pig too much during very hot weather because this could stress your pig and cause it to stop eating or lose weight. Cold weather can also present problems with stress. Try to use good judgment, and strive to keep your pig happy and eating well.

Ideally, you should practice walking or **driving** — the term for moving pigs — your pig every day for about a month before your show.

Grooming

Fitting your pig for the show ring also includes grooming your pig. You will need to wash your pig before the show. In order for your pig to really look its best, it is a good idea for you to wash it one or two times in the month before to the show. You will need to scrub and clean any stains or areas that are very dirty. You can use a mild detergent or a horse shampoo that you can find at a feed supply store, but it is best not to use a shampoo that is full of medications. White pigs will probably need extra cleaning, and you may have to use a whitening or brightening shampoo for bleaching to get the coat very white. You can also find shampoos for whitening and brightening the coat at feed supply stores or online from livestock supply stores. You can find very nice pig grooming supplies at a site like Vittetoe, Inc. (**www.vittetoe.com**). Use a stiff brush for washing and scrubbing. Pigs have thick skins, and it will not hurt them. Try not to get any water inside your pig's ear, or you may have a pig mutiny. Your pig will be very unhappy if you get water in its ears, and it will let you know it.

You should brush your pig daily for at least two months before your show. Brush in the same direction the hair grows to help it lie flat and smooth. This will train the hair to grow smoothly. It will also stimulate the skin and help the hair look more lustrous and healthy.

As the show approaches, you should check your pig's hooves. You will need to make sure they are neat and clean. In some cases they may be too long. If they are too long, you should ask someone with experience to trim them for you about two to three weeks prior to the show. Be careful not to trim them too short, or you could make your pig lame. Do not try to trim hooves when it is too close to the show date. You could make your pig lame if you accidentally cut too much hoof, and the pig would not be able to walk. You will need to use clippers to clip the long hair on the tail. Clip from the base of the tail to the switch on the pig's tail. Leave 3 to 4 inches of long hair on the switch of the tail. You should do this three to four days before the show. Trim around your pig's ears, and remove the long hair under the ear and around the edges.

If you are showing your pig during the winter and your pig has a long coat, you may need to use your clippers to trim your pig's underline so it will appear tidier underneath. Be careful not to get carried away or trim too much of your pig's coat beyond the underline.

There are also some things you should not do with regard to grooming. Do not try to "touch up" your pig or use chalk, powders, or other foreign substances if they are not allowed. Powders and oils are allowed in some states and not in others, and in some classes in some states and not in others. Make sure you know what is allowed before you try to use something. Your pig will be judged by knowledgeable hog people who know all the tricks to showing pigs. They will have their hands on your pig. It is best not to try to fool them with hair dressings or other substances if they are forbidden. Make your pig look its best and present it to your best ability. The judge will find the best animals.

Before you actually show, you will want to wash your pig the day before the show and spend extra time brushing and grooming your pig to make it look great. Make sure your pig's hooves are clean. Before going in the ring, you should check to make sure you have brushed off any sawdust or shavings from your pig.

The Show Experience

Many people are nervous at shows, but if you try to focus on your pig, it can help you stay calm. If you have prepared and practiced for the show while you have been getting your pig ready to show, then you should be ready to go in the ring. Showing your pig should be a fun and friendly experience because you are surrounded by other people who have the same interests as you. They care about farming, pigs, and the same kind of lifestyle. Try to enjoy yourself at the show, whether you take home a blue ribbon or not.

Shows are always publicized well in advance. It takes lots of planning and work to put on such large events. You can find entry forms in the show catalog or premium list. You should be able to find show catalogs and premium lists at feed stores and other places where agricultural products are sold. You can also check with your project leader if you are with FFA or a 4-H Club. They will make sure you get your entry in on time. Make sure you find out when the show begins accepting entries and when entries are due so you can enter early. Fill out the entry form carefully, and return it in plenty of time to make the due date. Entries for shows are modest, such as $10, plus the cost of ear tagging if that is required. Ear tagging is about $5. You will also need to plan for any parking fees on the day of the show.

You should receive a confirmation in the mail, or by e-mail, that your entry has been received, along with more information about the show. The show information you receive should tell you where the show ring is, what time you show, when to arrive with your pig and your supplies, and so on. If you have any questions that have not been answered in the material that is sent with your show confirmation, you should contact the show superintendent to ask. If you enter the show as a 4-H or FFA project, you should talk to your leader about how things will be arranged.

The show catalog or premium list should also contain the health requirements for the show. This will tell you what kind of vaccinations your pig may need before being allowed on the show grounds. You will need to make sure your pig meets these requirements and you have the paperwork to prove it before you take your pig to the show.

You will need to provide transportation for your pig to get to the show site. It is often best to travel with your pig at night or early in the morning, especially if it is during hot weather and you use a trailer or open-air truck. An open-air truck or a trailer can also be a problem in very cold weather. A closed trailer is better when the weather is very cold. It is a good idea to arrive with your pig at the show site the day before judging so your pig has time to adjust to the new surroundings.

What You Need at the Show

You will need to take enough feed with you to provide for your pig for the entire time you will be at the show. Do not change feeds while you are at the show because this may stress your pig, and you could bring on a case of diarrhea.

Find out if you need to take your own bedding to the show with you. Some shows provide bedding, and some shows expect you to bring your own. You can take shavings or straw if you need to supply your own. You will also need to take a tub or pan with you for feeding your pig. You will also need a water bucket and a shovel to clean your pig's pen.

You will need a tack or grooming box for your pig's grooming supplies and for some of your work things. It should contain the following:

⑤ Mild detergent or horse shampoo
⑤ Brushes

- Old clothes and boots to wear when you wash your pig
- Water hose
- Clean, soft cloths or rags for rubbing your pig
- Show stick or cane

Things to do at the Show

Once you have arrived at the show with your pig, you will need to find out where your pig is going to be housed. If you are with FFA or 4-H, you should check in with your leader for your pen assignment. Otherwise, you should ask the show superintendent for your pen assignment.

As soon as you know where your pig is going to be located, you can put down bedding and help your pig settle in. You will need to follow the show instructions for weighing in your pig, paint branding, and ear tagging your pig so it can be identified for the show.

Do not overfeed your pig at the show. Feed your pig lightly. Your pig should not appear shrunken while at the show, but you do not want your pig to look stuffed either. You should feed and water your pig outside the pen whenever possible. This can prevent your pig from turning over its water and getting the bedding wet. If you have to feed and water your pig inside the pen, then stay with your pig while it eats and drinks to keep it from turning the feed and water over. Make sure you continue to exercise your pig each day while it is at the show.

Presentation

You have prepared your pig for the show ring for months so you should be prepared when it is time to enter the ring. Be at the ring a few minutes early, ready to show. Allow yourself plenty of time to move your pig from the pen to the show ring.

You should wear appropriate clothes for your class. It is usual to wear a nice shirt or blouse, often of Western style, and a good pair of jeans. You can wear suitable shoes or boots that will protect your feet if a 200-pound pig with sharp hooves steps on them. Be sure your hair is neat. Braid it or pull it back if it is long. It is important to look neat and presentable. Your appearance can add to or detract from the impression your pig makes on the judge.

You may take a show stick or cane in the ring with you to drive your pig. Many people also take a brush with them into the ring. You should enter the ring on time when the class is called. Keep your eyes on the judge, and try to know where he or she is at all times. It is important to be aware and to pay attention. You will also need to watch what your pig does. Use your show stick or cane to move your pig by touching it behind the shoulder or on the side. Be sure you do not hit your pig too hard or strike your pig on the back or the rear. This is painful and unnecessary. You could startle your pig, and you will certainly make a bad impression on the judge.

Do not move too fast, and do not get caught up in a crowd of other pigs, especially in a corner. Ideally, you should try to keep your pig slightly away from other pigs so the judge can see it easily. Try to stay on the far side of the pig so the pig is always between you and the judge. This will give the judge the best view of your pig. If you walk on the judge's side of the pig, the judge may think you are trying to hide something from view. Try to remain at least 10 to 15 feet away from the judge. This gives the judge a better view of your pig.

You have practiced moving your pig and listened to comments about how your pig looks best. Try to show your pig to its best advantage. Most pigs look best when they walk instead of standing still so keep moving. Try to remember to keep smiling. If you act like you are relaxed, both you and your pig will feel better. The judge will respond to you better, too.

No matter how you place in the class, be a good sport. If you receive any kind of ribbon at all, you should shake the judge's hand and thank him or her. People do remember poor losers, and it leaves a very bad impression about you. On the other hand, if you win your class or place very well, be a gracious winner. Thank the judge, and be kind to the other exhibitors. Do not disparage anyone's pig at a show. Even if you think no one can hear what you say, it is surprising how often the things you say can come back to bite you. Remember that the pig that loses on one day may win at the next show.

After the Show

After the show, you will need to tend to your pig. If you have fed your pig lightly since its arrival at the show, you will need to get out the feed tub. Make sure you provide your pig with water because showing in the show ring can make a pig thirsty.

You will probably want to return to the show and watch some more judging. There is still a great deal for you to learn at any hog show, regardless of your experience. Before you leave, you will need to gather up all of the supplies you brought with you. Be sure you leave your pig's pen clean and pick up any trash.

Some shows will not allow pigs to leave until judging is completed. Check with your leader or with the show superintendent about what time you can leave the show site with your pig if you have any questions. Do be certain to make arrangements for taking your pig home with you the same night the show ends. Most shows do not make any allowances for pigs to stay after the show.

Summary

In this chapter, you learned about showing pigs. You have considered why people show pigs and how showing pigs can help you develop a good eye for livestock. You have examined what is meant by the term "type" in a breed. You have also considered 4-H and FFA and how they can help young adults develop good personal qualities and become better farmers. You have learned how to choose a good pig for the show ring and looked at what you need to do to raise a pig for showing. These things include proper feed, grooming, exercise, and training a pig to walk with you. Finally, you looked at your preparations for the show and the show itself. Showing pigs can be a wonderful experience, and you can learn a great deal. Many people enjoy showing pigs throughout their lives, or they may go on to become judges.

In the next chapter, you will examine what may be the end goal of all of your hard work: slaughtering your pigs. Whether you have one or two pigs you intend to have butchered for your own freezer or you have a larger number you intend to market, you will weigh your slaughter options. There are different regulations and inspections that apply to slaughter, depending on what you intend to do with your pork. You will also consider the different cuts of pork and ways to cure the meat. If you intend to sell your meat, you will need to consider packaging, and the next chapter will also look at what kind of packaging you should use. Finally, you will look at ways to market your pork.

This Little Piggie Went to Market

Up to this point, everything you have considered has been to raise your pigs with minimal stress so you could encourage your pigs to gain weight with the most efficient use of feed. You have considered how to choose your pigs; their housing and pasture needs; what kind of feed you might choose; and their health needs, among other things. After all of these months of work, your pigs should be at the right weight — normally 225 pounds — for you to make the decision to butcher them.

The decision to butcher your pigs may not be easy for you, especially if this is your first time raising pigs. Pigs have personalities, and you may have grown very fond of your pigs. But, you can try to remember that your pigs have a purpose, just like other farm animals. Your pigs are contributing to your own sustenance and survival, whether they go on your table or whether you sell them. It is appropriate to appreciate them and acknowledge the role they play in the agricultural way of life. One way or another, most people depend on farm animals for their food, though many people today no longer realize where their food comes from. Many people choose to raise pigs for their pork because they like knowing how their food was raised. Making the decision to butcher your pigs is part of the process.

Your Slaughter Options

You essentially have two options when it comes to slaughtering your pigs: You can do it yourself, which is not for everyone, or you can arrange to take your pigs to a butcher and have him or her do the job. Naturally, there are some variations on these options, but these are the basic choices.

There are no regulations governing home slaughtering. If you decide to butcher your pigs yourself, you are free to kill your pigs and butcher them as you prefer. Your pigs do not have to be inspected if you do not intend to sell any of the meat. The meat must only be used for your home consumption, and you cannot sell any of the meat. But, if you intend to sell any of the pork from your pigs, your pigs must be slaughtered and processed by a licensed and inspected processing facility. Who butchers and processes your pigs will depend on how you intend to market and sell your pork. Licensing and rating for butchers ranges from custom, state-inspected, USDA-inspected, to organic. There does not seem to be one central listing for butchers in the United States or in each state, but you can search the Internet for butchers in your area or check your local telephone listings.

⑥ A custom butcher is only licensed to butcher and process animals that will be consumed by the person who owns the animal. The meat cannot be sold to consumers. When the pig is butchered, the meat is packaged and labeled as "Not For Sale." Custom processing facilities are inspected and required to meet sanitary conditions and some of the same labeling and storage requirements of the USDA.

⑥ The rules for state-inspected facilities vary somewhat from state to state but, generally, meat butchered and processed by a state-inspected facility can only be packaged and sold inside that state. As you might guess, these facilities are inspected by inspectors from a state agency. About half of the states in the United States have this kind of state inspection system in place. The state inspection system must be "at least equal to" the regulations and guidelines provided by the USDA.

⑥ If you wish to have your pigs butchered and processed at a USDA-inspected facility, then the animals will need to be inspected both before and after butchering by a USDA-certified inspector. USDA inspection does cost more, but you will be able to sell your meat anywhere in the United States. There is a per-animal fee, as well as any additional costs from extra package labeling, but the additional marketing opportunities often make USDA inspection worth the cost. Many restaurants and grocery stores prefer to purchase meats that have the USDA seal of approval.

Processing fees, which includes killing your pigs, can cost from around $20 to $60 per animal for large hogs. There is also a processing fee for the dressed weight of the pig. This fee is around 30 to 45 cents per pound. There may be further fees for more processing, such as curing, boning, or smoking the pork.

There are currently very few organic butchering and processing facilities in the United States. Some of them can process both organically raised and non-organically raised pigs. However, the processing tools and all of the equipment used must be cleaned between processing the organic and non-organic meats in order to keep the organic meat separate. Prices for processing at organic facilities can be much higher. When pork is processed to meet organic standards, it cannot contain any synthetic artificial ingredients, additives, or preservatives. There is typically minimal processing. The materials used for packaging cannot contain synthetic fungicides or preservatives. And, there must be specific labeling that identifies the meat as organic according to USDA labeling requirements.

According to the USDA, no claims can be accurately made on the labeling that organic pork is in some way inherently better than pork produced by other, more traditional methods of production. If you are interested in finding an organic processor in your area, the best way to find one — other than talking to other organic growers — is to contact your state agriculture department. They should have a listing of organic processors if there are any in your state. LocalHarvest (**www.localharvest.org**) is also a good website for more information about finding more organic meat processors.

On-farm processing is also an option in a few places in the United States. On-farm processing involves using a mobile processing unit (MPU). These units have sprung up in areas where there is a lack of processing facilities. Most MPUs are state licensed, though some may be USDA-inspected. MPUs are only able to process a few animals so they are not a good option if you have a lot of pigs to butcher. If you have an MPU in your area and you choose to use it, you should be aware that they often expect the farmer to do much of the work him or herself or to help with the slaughtering, packaging, and clean up. But, if you do not have a butcher or processing facility located near you, then a mobile processing unit may be something to consider if you do not want to slaughter your own pigs.

Making Arrangements With a Butcher

If you raise just a few pigs of your own for home consumption, then autumn is the traditional time for butchering your pigs. This way, you will not have to be concerned about feeding and housing your pigs over the cold winter months. You will have plenty of bacon and other pork in your freezer for the next year, and you can buy a few more weaner pigs in the spring to start the process over again if you like. This is often a very economical way to raise pigs and feed your family with a minimum of expense and work.

On the other hand, if you have your own herd and you are breeding several litters per year so you can sell the pigs to others or sell the pork, then there is no problem with making arrangements with a butcher to kill and process your pigs at various times during the year as your animals reach their ideal slaughter weight. Depending on the time of year, butchers may be in high demand so you should make your appointment with them well in advance. Hunters also use butchers for processing and dressing deer meat and other game, for example, and some butchers work with many different kinds of livestock.

If you have a choice of butchers, you should make arrangements to visit them in advance to see their facilities and talk to them. Find out how they are licensed and inspected, as well as what they charge. You will need to make sure they meet the right inspection criteria for the selling method you have in mind. For instance, if you intend to sell your pork products nationally, do not take your pigs to a state-inspected butcher.

You should also ask to purchase some samples so you can see how the butcher prepares the meats. Most butchers have their own recipes using different spices for making sausages and for curing meats. There can be a wide range in the amount of salt or smoke used in curing meats and in the kinds of spices used. You will want to make sure you like the way these pork products taste before you commit to allowing a butcher to process your pigs. Pay special attention to the ham, sausage, and bacon the butcher produces because these products will be most popular with customers. If you do not like their flavor, then your customers may not like them either. Anything produced will reflect on you and not the butcher so make sure you like what the butcher will do to the meat.

CASE STUDY

Ben Chamberlain
Sleepy Panton Farms
186 Spaulding Road
Panton VT 05491
sleepypantonfarms@gmail.com
www.sleepypantonfarms.com
802-475-2393

Ben Chamberlain of Sleepy Panton Farms in Vermont breeds and raises heritage pigs. He has raised pigs for two years and has 12 pigs of his own. He also works at a pig farm that has 54 sows and eight boars. They raise hundreds of piglets and feeder pigs. The pigs are pasture-raised Berkshires and Tamworths.

Chamberlain said he raises pigs because he enjoys them, and he likes being outside. "I like to cure and smoke pork bellies, raise my own food." He also believes that it is important to know how food is raised, as well as what its living conditions are. He likes the idea of bringing local, sustainable agriculture into his own backyard.

Chamberlain said he started raising pigs with two meat pigs because pigs hate to be alone. "Once I got the hang of it, we bought a baby Berkshire boar and a young Tamworth gilt. Now, we're breeding." Chamberlain said he crosses the Berks and Tams to get bigger hams, beautiful pork bellies, and better flavoring.

His pigs are primarily raised outdoors in the woods and field. He has two pig "bunkers" set up for them to sleep. "Here in Vermont the winters are cold so they have the choice to sleep or give birth in the barn or a three-season covered deck. The benefits are huge! Pigs need room to roam and dirt to dig."

Chamberlain says the most important things to know about raising pigs are:

- Pigs need room to roam.
- Pigs need other pigs; they are highly intelligent and social animals.
- Pigs grow quite large if you let them. Be ready.

Chamberlain does have a favorite story about raising pigs. "The first year I had pigs I learned that because pigs can't sweat, they can overheat. As I looked into the scorching midsummer sun, and then at the pen, I realized there was no pond. So, I started shoveling out one, and the pigs eagerly joined — pushing more dirt with their noses than I was with the shovel. In no time, we had a 10-foot-by-10-foot-by-2-foot deep hole in the low point of the property. As I started to fill it with the hose, the pigs began dashing through the jet stream jumping, grunting, trying to bite the water in the air and rolling around together. It showed me these animals have such personalities, emotions, and thoughts."

Finally, Chamberlain says that owning pigs is highly addictive.

Some butchers may allow you to be on-site while a pig is processed so you can observe the procedure if you are comfortable being present. If you are present during the entire procedure, you can tell if there is anything you find objectionable or if the butcher's work is acceptable.

The butcher's processing will normally include cutting your pig into standard cuts of meat and curing the hams, sausage, and bacon. If you would like the meat prepared some other way, such as cut in different cuts or leaving some of the meat uncured or "green" so you can cure it yourself at home using your own recipes, you will need to be sure to discuss this with the butcher prior to butchering and processing. Make sure it is fully understood and agreed upon before the butcher begins work on your pigs. You will need to pick up these fresh — uncured — meats as soon as the carcass is chilled and ready to be cut or cured. Remember, however, that you will not be able to sell these home-cured or home-smoked meats. They can only be for home consumption because they were not completely processed by an inspected butcher.

When discussing arrangements with the butcher, you should also discuss packaging. Packaging can make a difference to customers so find out what kind of packaging the butcher normally uses. White butcher paper wrapped around the meat will be fine if you are having the meat processed for home consumption. But, if you plan to market and sell your meat, it is best to choose clear, vacuum-wrapped packaging. Cryovac®- or vacuum-sealed packaging costs more to produce, but it allows the consumer to see the meat, which can make it more appealing to consumers. Being able to see the meat is also helpful when the meat is in the freezer at home. You will pay more for this kind of packaging, but it may be worthwhile. Talk to the butcher about packaging options, and see what he or she has available and what they cost.

Getting your Pigs Ready for Slaughter

Once you have chosen a butcher and made an appointment a month or so ahead of time, you will need to get your pigs ready for slaughter. The ideal weight for slaughter is around 225 pounds. Any weight your pig adds more than 225 pounds is usually deposited in the form of fat. Unless you are trying to raise pigs for their lard, it is a waste of feed and money to keep feeding a pig after it weighs 225 pounds.

You can calculate when your pig will be ready for slaughter by determining its weight. A pig that weighs 150 pounds will gain about 1.8 pounds per day between 150 pounds and 225 pounds. That means that when it weighs 190 pounds, it will take about 20 more days for it to weigh 225 pounds.

There is an easy and generally accurate method for determining how much your pig weighs without putting it on a scale. You can measure the **heart girth** of your pig in inches — the distance around the pig just behind its elbows — and then, measure your pig's length, from between the ears to the base of the tail. Use the following calculation:

> Heart girth x heart girth x length, divided by 400 = estimated weight

Once you know your pig's approximate weight, you should be able to see if you are on track for your pig to weigh close to the right amount at the time scheduled for butchering. You may increase feed in the last few days, if necessary.

If your pig is going to weigh a little more than 225 pounds at the time of processing, there is no need to panic. Consumer tastes tend to prefer leaner meats these days, but some excess fat can be trimmed off. If the meat is good, people will appreciate it. Some breeds are known for having more fat as marbling, and the fat adds taste to the meat. The important thing is to know your breed and know what is ideal for it.

If your pigs are being given antibiotics in their feed, or other supplements that should not be given to humans, you will need to discontinue feeding them to your pigs during these last several weeks. You should consult the labels on anything you are giving to your pigs and make sure you follow the label directions to discontinue use prior to slaughtering. There is often a withdrawal period of several weeks between the time an animal stops receiving a medication or supplement and when it is butchered. Be sure you comply with these safety instructions for human consumption.

As the time for slaughtering and processing approaches, you will need to arrange transportation to the butcher for your pigs. Whether you have one or two pigs or a much larger number, it is very important to keep your pigs calm and stress-free before, during, and after transport. Pigs that are stressed and upset prior to being butchered do not have very tasty pork. You do not want to do anything to cause your pigs to release too much adrenaline into their bloodstream from becoming anxious.

If you have the space, you can place the pigs headed for slaughter in a separate pen a couple of days prior to their trip to the butcher. You can wash them down at this time. Most farmers also remove feed from pigs 24 hours prior to taking their pigs to slaughter. This reduces the chance of contamination from food being digested in the animal's digestive tract. It also makes it easier to remove the internal organs of the pig.

Ideally, you can use a livestock trailer to transport your pigs to be processed. Make sure it is comfortable for your pigs. It is best to load the pigs into the trailer the night before you intend to take them to the facility so they have plenty of time to calm down and relax in the trailer. Do not crowd too many pigs into the space because this will make them anxious, and they could become aggressive toward each other. If the weather is warm, you should be careful to park the trailer in the shade. Make sure you provide the pigs with plenty of water. They should have all the water they want while in the trailer. It is important that the pigs are not dehydrated before butchering because this will affect the final pork products.

It is recommended that you transport pigs together that already know each other. If you attempt to transport your pigs with pigs they do not know, you will likely have fighting, and pigs that are very upset. This is the last thing you want. You should also avoid transporting two unfamiliar boars at the same time, as they are likely to fight.

Keep your pigs calm overnight. Once you are ready to leave for the butcher in the morning, your pigs will probably lie down in the trailer once you begin driving. Drive slowly and carefully. Once at the facility, you should move the pigs slowly, without yelling at them or trying to make them move faster. Keep them calm. At the processing plant, the butcher and his or her helpers will take over.

Cuts

Whether a butcher is processing your pig for you or you are slaughtering your pig yourself, you will need to know about the cuts of meat that can be obtained from your pig. The pig is one of the most bountiful of all animals, and people often say you can use everything but the "oink" when it comes to preparing a pig.

There are four basic or primal cuts of fresh pork that are commonly made: the shoulder, the side, the loin, and the leg. You can obtain the following meats from these cuts:

Shoulder	Shoulder butt, roast, or steak
	Blade steak
	Boneless-blade Boston roast
	Smoked arm picnic
	Smoked hock
	Ground pork for sausage
Side	Spare ribs/back ribs
	Bacon

Loin	Boneless whole loin (butterfly chop)
	Loin roast
	Tenderloin
	Sirloin roast
	Country style ribs
	Chops
Leg	Ham/fresh or smoked or cured
*Source United States Department of Agriculture	

When talking to your butcher, you can discuss these cuts. Your butcher will probably have a processing sheet so you can check off which cuts you want to have made from your pig. If you butcher your pig yourself, it will take some practice to produce all of these cuts just the way you want them, with the right thickness, trim, and other qualities. You may also have to experiment with recipes for curing to get the taste of the cured meats to your liking.

Other parts of the pig are also used. Pig's ears, brains, kidneys, and other organs, as well as pig's feet and the pig's tail, are frequently found on the menus of chic restaurants these days. Chitlins or chitterlings are an old-time Southern favorite. They are pig intestines and can be served either as a stew or fried. Cracklins are fried pork rinds, or pig skin. The fat from your pig can be used as lard. You can use virtually everything from your pig in the kitchen.

A pig that weighs 225 pounds will present you with about 75 percent of its body weight as a dressing percentage, or 170 pounds of meat, bones, and fat after slaughter for the carcass or hanging weight. You should figure on being able to use about 60 percent, or 102 pounds, of this weight as pork you can eat or sell. The biggest part of this usable pork will be the ham, which usually accounts for about 23 percent of the carcass, or about 23 pounds in this case. The side and the loin areas will each make up about 15 percent of the carcass, or about 15 pounds each. The picnic and the Boston Butt from the shoulder will each account for about 10 percent of the carcass weight, or 10 pounds each; and the miscellaneous parts, such as the feet, the jowls, the skin, the fat, and the shrink, or the amount of weight loss because of urination and defecation, will account for about 25 percent of the carcass weight, or 25 pounds in this case.

> ## So, to sum up, 225 pounds live pig will result in:
>
> - 170 pound carcass or hanging weight breaks down to:
> - Ham: 23 percent
> - Side cut: 15 percent
> - Loin cut: 15 percent
> - Shoulder cuts: (picnic and Boston butt) 20 percent
> - Miscellaneous parts: 25 percent
> - Total retail pork weight: approximately 102 pounds

There will be some small variation in the amount of lean pork and fat from your pigs depending on whether they are gilts or barrows. Gilts produce more lean pork than carcasses from barrows of the same weight.

After Processing

After processing, you will need to pick up your meat from the butcher within a day or two, depending on the butcher's refrigeration storage capacity. Naturally, you will need to make sure you take iced freezer chests with you to transport the pork, particularly the uncured pork, such as shoulder and loin cuts and any miscellaneous parts. You will need to have good freezer storage capabilities at home in order to store the pork, especially if you have pork from several pigs and you intend to sell some of it.

If you will sell some of the pork for retail sale, you should carefully inspect the packaging at the butcher's facility before taking it home to make sure the packaging is what you requested.

Slaughtering at Home

Slaughtering your pigs yourself at home is not for the tenderhearted, but if you wish to control every aspect of raising your pigs and producing your pork, it may be something you want to do. But, before deciding you want to slaughter your pigs or engage in home butchering, you should thoroughly consider the task. Killing large animals is no easy task, no matter what method you use. You will need to have the physical strength to hoist and move the heavy, dead weight of your pigs. Once you begin the job, you will need to finish it because the meat will quickly start to spoil if you delay. You will also need to work with a number of sharp knives and other special implements during butchering. If any of this gives you pause, you may wish to send your pigs to a good butcher instead of trying to do this job at home. Most people who raise pigs do use the services of butchers.

If you are still interested in the idea of slaughtering your own pigs, then it is recommended that you visit someone else's farm when they are slaughtering pigs so you can see first-hand what is involved in slaughtering and butchering pigs at home. You may be able to assist and get some experience before you attempt to do this job with your own pigs.

Planning ahead

In order to successfully slaughter your pigs at home, you will need to do a great deal of planning ahead. If you will be slaughtering at home, then you will be much more dependent on the weather than if you were using a butcher. You probably will not have access to the temperature-controlled environment that a butcher has. That means you will most likely need to slaughter your pigs in the cooler months, or at least during the very coolest part of the day, such as the early morning hours. Alternatively, you may be able to do the slaughtering yourself and arrange with a local butcher to chill and cut the carcass for you. If you do use a local butcher to chill and cut the carcass, you will need to make the arrangements far in advance for the same reasons that someone taking their pigs to the butcher needs to make arrangements a month ahead of time: Butchers can be in high demand at some times of the year so book ahead of time if you would like them to chill and cut your meat, even if you intend to do the actual dispatching yourself.

If the temperature is less than 30 degrees, you can slaughter pigs at any time because bacteria that could spoil the meat will not grow fast at these temperatures. But, if the weather is extremely cold — below the mid-20s — you should not allow the carcass to freeze right after slaughter because this will make the meat less tender than if it is chilled first before freezing. You will also need to carefully select where you will slaughter your pigs. Your space requirements will depend, to some extent, on the methods you intend to use. After killing your pig, for example, you will need to remove the animal's skin and hair. There are two ways to do this. You can either scald the pig by placing it in a large vat of water, or you can skin the pig. If you choose to scald the pig, you will need access to a large vat of water, some way to heat the water to boiling, as well as a way to hoist the pig up over the vat and lower it into the water. This can be accomplished by using a tree limb and swinging the pig over it using meat hooks with assistance from others, and so on. You can move the pig with help from others to lower it into the vat of scalding water and raise it several times. You will need to be able to apply chains to your pig's hind legs in order to raise it over a tree limb, for example. It is easiest to do this if you can attach the chains to a vehicle to help lift the pig off the ground and raise the chains over the tree limb.

You will also need to have a place to slaughter your pigs. This can be done outdoors, but it is often better if you can do it indoors to keep dust and debris away. You will need to drain blood from your pig at this time so the site should be a place that is easy to clean up. It is often easiest to do this right on the ground, especially if you have a concrete floor or if you place a tarp on the floor. Others prefer to slaughter the pig outdoors. You can put down plenty of straw to help soak up some of the blood on the ground. Wherever you choose to slaughter your pigs, you should clean up the site before you begin your work. The area should be clean and sanitary. If you are working outdoors, clear the area of leaves and debris so they will not blow up on to the carcass while you work. Place a layer of straw on the ground where the pig will be suspended and the blood is to be let. If you are working indoors and the area has a wooden or concrete floor, wash the floor and all of the equipment with soap and water before you begin. Make sure you rinse the area completely because any sanitizers could discolor your meat or alter the meat's flavor.

It is very important that you have good, sharp tools for slaughter. You will need sticking knives for sticking the pig to let the blood drain, skinning knives for skinning the pig, boning knives and butcher knives for cutting meat, a steel sharpener to keep your blades sharp, meat saws to cut through large sections, and meat hooks for moving large hunks of meat. You can find these tools at a place such as HomeButcher.com (**www.homebutcher.com**), among others. Tools are not cheap. A skinning knife will be around $25. A boning knife is $16 to $19, depending on which kind you get. A manual meat saw is about $50. Other useful implements to have on hand as you prepare the pork cuts include:

- Thermometers
- A meat grinder
- Meat needles so you can sew rolled cuts of meat
- Hair scrapers to scrape the hair off the pig
- Hand washtubs
- Clean dry towels
- Soap
- Vats for hot and cold water

A manual meat grinder can cost anywhere from $50 to $90, while electric meat grinders range from $100 to $2,800. All of your tools should be thoroughly cleaned before use with dish soap. You will also need a sturdy table to hold all of your tools. You can also purchase pre-blended seasonings and cures at a place like HomeButcher.com if you do not want to use your own recipes.

If you have the proper tools for slaughter, it will make your job easier, and your work will be more efficient. You should plan on your first pig slaughter taking about two to three hours, which is much longer than it will take you after you become more proficient.

Prior to slaughtering pigs at home, you should follow the same procedures you would follow before taking your pigs to a butcher. You should remove your pig to a holding pen a day or two prior to the slaughter date. You should withdraw feed 24 hours prior to slaughter but keep water available continuously to avoid dehydration. You should keep your pigs calm and stress-free during this time.

Before slaughtering day, you should study up on pig anatomy. Make sure you are familiar with where the pig's organs are, where the bones are, as well as the pig's digestive system. You will be seeing them all up close very soon. You should also make sure you know where the jugular veins are so you can make a good cut in the neck.

Before the day for slaughter, you should try to round up some people who can assist you. Simply having people on hand to help you move the carcass can be a great help. If you have people to help you who have experience with butchering, so much the better.

Killing your pig

There are several acceptable ways to kill a pig for butchering. The fastest, and perhaps the most humane way, to kill your pig is by using a sticking knife to quickly and efficiently stick your pig in the jugular vein in its throat. This is easiest to do if you have the pig raised above you with its head hanging down. You can do this if you have placed chains or straps between the pig's hocks and hooves and lifted it by raising the chains over a tree limb or over a beam in a building. The pig will be very unhappy about this position but you should be able to press your sticking knife against the point of the breast bone and make a 4-inch vertical incision in the middle of the neck. This cut should sever the jugular vein and the blood should begin to flow. You should have tubs placed below the pig to catch the blood to prevent the area from becoming too messy. If the area is too slippery, you will not be able to work properly. When pressing the sticking knife into the breast, you should be careful not to press too deeply or you could cause internal bleeding that can affect the meat.

Butchers will normally use an electrical stunner to stun the pig before cutting the throat. On some farms the farmer will use a mechanical stunner to stun the pig first. A mechanical stunner can be used anywhere, whereas an electrical stunner needs electricity to operate. Many farmers will not have access to electricity if they kill their pigs in a field. The electrical stunner is used more often by a butcher when slaughtering large numbers of animals. Other people prefer to use a .22 caliber rifle to shoot the pig in the forehead between and slightly above the eyes. With any of these methods, it is absolutely essential that you are accurate and do not cause the animal any distress. The kill should be swift, and you should not allow the animal to suffer. You should try to stay very calm prior to making the kill. If you are upset or agitated, then it will also upset the pig. Obviously, if you use firearms or stunners, you should take all necessary safety precautions.

No matter which method you use for killing the pig, the next step in the process is to bleed the pig. This needs to begin within about two minutes of killing the pig so the blood will flow freely. Pigs that are hung upside down do bleed the best, but if you do not have a way to suspend the pig, you can bleed the pig with it lying on the ground on its back. If the pig is stunned, you can have someone stand over the pig on the ground and hold its front legs. Locate the edge of the breast bone and thrust your sticking knife under the bone with the sharp point aimed toward the tail, and then thrust upward to sever the carotid artery. The pig should bleed, but if it does not, you can insert the knife a little more deeply, and there should be plenty of blood. Again, you should have pans ready to catch the blood.

Removing hair and skin

In order to remove the hair and skin from a pig, they have traditionally been scalded in hot water and scraped with the skin left intact. However, today, many people prefer to skin the pig because it is easier and requires less equipment. It will be up to you to decide which method of removing hair and skin you prefer to use. Many chefs and food aficionados today find uses for the skin of the pig in food dishes. Whichever method you prefer, you can find uses for the skin if you are interested in keeping and reusing it.

SIDEBAR:

In case you are wondering, footballs in the United States are not actually made from pigskins. They are made of cow leather, though at one time soccer balls and other similar sports balls are said to have been made from the pig's bladder, giving rise to the term "pigskin." The skin of the pig does make a very soft, durable leather, and it is used today to make gloves, clothing, and other accessories. Pigskin has been used to make fine saddles for a very long time.

Scalding and scraping

The purpose of scalding the pig is to loosen the pig's hair and the **scurf**, or the layer of skin oil, dirt, and cells on the skin, so they can be more easily removed. It has long been believed that it was necessary to leave the skin of the pig intact on the pig in order to achieve proper curing for the ham and bacon. But, in recent years, more people tend to skin the pig and cure pigs without the skin. If the pig is skinned badly, it can ruin your bacon, though. Scalding the pig requires more work and more equipment than skinning the pig.

In order to scald the pig, you will need a heat source and a water source. Most people use 55-gallon drums or barrels. You should start by heating approximately 50 gallons of water close to boiling while you are killing the pig and letting it bleed. This water should be either heated in the barrel you intend to use for the pig or transferred to the barrel when you are ready to place the pig in it.

You will likely need help moving the pig into the barrel. Some people prefer to dig a shallow hole for the barrel that will contain the pig so it is easier to place the pig inside the barrel. If you dig the hole at a slight angle, you can kill and bleed out the pig next to the vat or barrel, and it will be easier to move the pig into the vat when you are ready to place the pig into it. You should make sure you do not make the angle too low, or the vat will not contain enough water to cover the pig.

Another method is to build a fire beneath the vat you will use to scald the pig. In order to use this method, you will need to dig a pit for the fire and rig a sturdy method for suspending the vat over the fire while the pig is in the vat. If you can place heavy metal legs on the vat, such as a cauldron, then it could hold the weight of the pig over the fire.

You will also need to use a thermometer to assess the temperature in your vat. It is best to scald the pig slowly at a temperature around 140 degrees. At this temperature, it will take between three and six minutes to scald the hair and scurf from the pig. The pig's hair can be very difficult to remove, especially in the fall when the pig has begun to grow thicker hair for winter. If you scald your pig during the fall and your pig does have this winter growth of hair, you may need to use higher temperatures, between 146 and 150 degrees, or keep your pig in the scalding water for longer periods. You may also wish to add ¼ cup of rosin, lime, or another alkaline mix to the scalding water to help remove the scurf. This will also make your pig's skin appear whiter.

It can be difficult to keep the water in the vat at a precise temperature. You will need to continue to check the temperature of the water in the vat by using your thermometer throughout the process. You should have more boiling water ready to add if necessary. You can also add cooler water if the water is too hot. If you begin with a water temperature between 155 and 160 degrees, the water should be at a scalding temperature when you are ready to place the pig in the vat. If it is cold weather, the cooler outside temperature will cool the water in the vat faster.

Once the pig is in the vat, you will need to keep it moving and pull it from the barrel several times in order to keep it from overscalding. If the pig begins to overscald, it will cause the skin to contract around the base of the hairs, known as "setting the hair." This effectively cooks the skin of the pig. If the skin becomes overscalded, it will make the hair very difficult to remove.

When the water reaches the correct water temperature of 140 degrees, you should place the pig in the vat head first. Turn the pig in the vat, rotating it, and pull it in and out of the water occasionally. You should check the pig's skin often for signs that the hair is easy to remove. The hair should start coming off first over the back and sides and then in the flank areas. After you can remove the hair easily from the flanks, you should remove the pig from the vat and place the pig rear first into the vat. Be sure to check the water temperature, and raise the temperature back up to 140 degrees. The temperature will no doubt have cooled during the several minutes that the pig has spent in the vat.

Once the rear of the pig is in the scalding water in the vat, you can start pulling the toenails and dew claws from the pig's front feet. You can insert a hook into the top of the nail and pull in order to remove the nail. Start scraping as much of the hair off the head as possible, paying special attention to the hair around the ears and snout. You can use a knife or a bell scraper for this job, which is a scraper made for removing hair. Some bell scrapers have a hook on the end for removing the toenails.

You should continue to turn the pig in the barrel so it does not overscald. Lift the pig out of the barrel, and test the looseness of the hair. Once all of the hair is loose and easy to remove, you can remove the pig from the barrel.

Ideally, you will have a sturdy table to place the pig on so you can continue to remove the hair. Alternatively, you may place the pig on a piece of plywood on the ground to continue working. Remove the nails and dew claws from the pig's back feet, and remove the hair from the pig's tail. Work to remove the hair on the pig's legs by gripping and twisting the hair. Work on the difficult areas, such as the head, feet, and jowl first, while the pig's skin is still hot from the vat. Then, you can move to the easier areas, such as the pig's back and sides. You can use the bell scraper for these easy areas. Try tilting the scraper up toward the forward edge and pulling it forward, using as much pressure as possible.

You will need to work quickly because the pig's skin will set as it cools down, making it harder to remove the hair. If you find areas of hair and scurf that are hard to scrape, you can cover them with a piece of burlap and pour hot water over the material to loosen them. You can make scraping the legs and head easier by moving them if they begin to set. This will keep the skin stretched and loosened.

After you have removed most of the hair, you will need to pour hot water over the pig so you can continue to scrape. You should place the bell scraper against the pig's skin and move it in a rotary motion. This will help remove the scurf and the remaining hair. If there are still patches of hair that are not removed by the bell scraper, you can use a knife. If you prefer, you can use a knife for the entire skinning process. You can use whichever tool you are more comfortable using.

At this point, you are ready to hang and suspend the carcass. You can start working on the carcass by cutting off the soles of the feet. Cut between and around the toes. You should use your knife to expose the pig's gambrel tendons. You can do this by cutting through the skin that lies on the back of the rear legs from the dew claws to the hocks. You should carefully cut along each side of the tendons. Be very careful not to cut the actual tendons, otherwise you will have no way to suspend the pig. You should insert the spreader or gambrel — the instrument called the gambrel — under the tendons in each leg in order to expose them. Fasten the legs to the spreader bar, and suspend the carcass from it. The legs will need to be spread at least 14 inches apart. When your pig is suspended, you should make sure neither its head nor its forelegs touch the ground to avoid contaminating the carcass with bacteria.

You can use a blowtorch or a small propane torch to singe off any remaining hair and scurf from the pig. Singeing will remove most of the remaining hair, and it will darken other small hairs so you can see them. You will need to be careful that you do not burn the skin when using one of these torches. Burning the skin can not only make the skin unsightly, but it can also affect the flavor of the meat. You do not want to partially cook the pork at this point in your processing. You can shave off any remaining hair and wash the carcass completely.

Skinning

It takes less time to skin a pig than to go through the scalding and scraping procedure. It also takes less equipment. You will still need to be able to hoist the pig up to work on it, however. After you have stunned and bled the pig, you can place the pig on a sheet of plywood, some concrete, or in some straw. You should wash the blood and any dirt or debris from the carcass. Turn the pig over onto its back, and hold it in position with blocks, such as cinder blocks, on each side.

You can begin the skinning process by cutting the skin around the pig's rear legs, just below the dew claws. You should make a cut through the hide and down the back of the leg. Continue to cut over the hocks and up to the midline at the center of the hams. You should carefully skin around each side of the leg. Remove the skin to a point below the hock.

At this point, you can open the skin down the animal's midline. Cut from the point where the pig was stuck, around each side of the genitals, and move on to the anus. You should make this cut by inserting the point of the knife under the skin with the blade held upward. This process is called cutting from inside out. It protects the meat from becoming contaminated from materials on the hide. You should avoid cutting too deeply because this could puncture the intestines and contaminate the meat.

Next, you should remove the hide from the inside of the hams. You will need to be careful because it is easy to cut through the fat layers and into the lean meat. You must continue to be careful not to contaminate the meat. You should continue to skin along the animal's sides toward the breast area. You should grip the loosened hide in the opposite hand and pull it up and out away from the animal. By doing this, you will produce tension in the hide, which will remove any wrinkling. It will also allow the knife to slide smoothly and easily. Hold the knife firmly, and place it against the hide, turning the blade slightly outward. You should skin as far down the sides of the animal as possible, but do not skin around the front legs. The skin around the front legs will be removed separately.

You should return to the rear part of the carcass and remove the hide left on the rear of the hams. You should not skin the outer portions of the hams at this point. Wait until you have raised the carcass to a better working height. You can remove the pig's rear feet by sawing through the bone. Cut about 2 inches above the hock. You should insert the spreader or gambrel under the large tendons on the rear legs and attach the legs to the spreader securely.

You can hoist the carcass to a convenient working height to remove the skin from the outside of the hams. This is about waist high. Start skinning around the outer parts of the hams, but leave as much fat as possible on the carcass. Next, you should remove the hide from around the anus and cut through the tail at the joint closest to the body. The tail is edible and some people like to save it. Continue and pull the animal's hide down over its hips. You should be able to pull the hide from the hips and back off and slide it over the hips. This will leave layers of fat on these sections of the carcass. You may need to use a knife to cut between the skin and the fat in some cases if large sections of fat are coming off when you remove the hide. Some fat needs to remain on the pig to protect the meat inside, to give the pork flavor later when it is cooked and to cure some of the cuts. You will be able to decide during the meat cutting process how much fat you want to leave on each cut.

You should hoist the carcass to its fully raised position. Cut open the hide down the back of the forelegs, and remove the hide on each side of the forelegs. Proceed to skin along the inside of the forelegs and the neck. Move to the shoulders and jowls and then to a halfway point to the back of the carcass, skinning as you move along. You can slowly pull down and out on the hide and remove it from the animal's back. The skin should come off easily, leaving the fat beneath it intact, but if the fat does begin to tear, you can use a knife to pat it back into place. Continue pulling the hide as far down the back as possible. When it become difficult to remove the hide around the animal's neck, you can complete the removal with your knife.

If you intend to save the pig's head, you will need to skin over the animal's head and down the face, cutting at the snout. You should remove the front feet by sawing just below the knee joint.

Evisceration

Evisceration refers to removing the pig's organs and intestines. During the evisceration process, you will need to be very careful not to allow the contents of the stomach, the intestines, or the other organs to come in contact with the carcass and meat. Some of these organs contain bacteria and fermenting food that can harm your meat and cause it to spoil. You will also be removing the **leaf fat**, or the fat around the pig's kidneys, that is used for making lard. The pig's head will also be removed. By the time you are finished with this part of the slaughter process, only the meat sections of the carcass will remain.

Your pig should be hanging, head down, from the gambrel. It is a good idea to have a large tub positioned under the pig in order to catch the intestines and other organs as they come out.

You will begin the evisceration process by removing the pig's anus. You do this by cutting around the anus and making a cut deep into the pelvic canal. Pull the anus outward, and cut any remaining attachments to it. Be very careful not to cut the large intestine as its contents could contaminant the carcass. Once the anus is loosened, you should tie it off with a piece of string so it will not cause any contamination.

When slaughtering a barrow, you will need to remove the penis or "pizzle." You should cut through the skin and the fatty tissue that lies along each side of the penis and around the opening. Lift the penis upward and make a cut beneath it along the midline. You should cut along the penis and between the hams, and then pull the penis upward, and remove it where it attaches at the base of the ham. There is a natural separation between the hams. You can continue the cut you have made between the hams and expose the white connective tissue. Cut through this connective tissue to the pelvic bone or aitchbone. Continue to cut through the cartilage between the pelvic bone and separate the two hams. You should be able to do this in young pigs; however, you may need to use a handsaw to split the pelvic bone in older pigs.

Next, you can proceed to open up the pig's chest cavity. You can make a cut from the place where the pig was stuck to the upper end of the sternum or breastbone. Insert your knife at the top edge of the sternum, and make a downward cut. It is best to cut slightly off center. Proceed to open up the chest cavity.

Next, you should open the animal's midline. Begin at the opening you made to split the pelvic bone. Insert the handle of the knife in the opening, and point the blade outward, away from the inside of the pig, in order to avoid cutting the intestines. Open the midline all the way up to the breast opening. Let the intestines and the stomach roll out and hang. Do not let them fall out because they are still connected to the esophagus. If they tear at this point they will spill their contents into the carcass and contaminate the meat. Proceed to pull the loosened large intestine past the kidneys. Cut the connections to the liver, and remove it by pulling outward on it and cutting the connective tissue. Next, you can remove the gall bladder from the liver. Simply cut beneath it and pull. Do not allow the contents of the gall bladder to spill onto the liver.

Carefully pull the stomach and intestines out, and cut through the diaphragm. The diaphragm is a thin sheet of muscle and white connective tissue. It separates the stomach and the intestines from the lungs and heart. Next, you should pull outward on the lungs and the heart, and cut down along the sides of the windpipe to cut its attachment at the head. You will also need to separate the heart from the lungs. You can do this by cutting across the top of the heart. You should split the heart open so it can be thoroughly washed. You should wash the heart and liver thoroughly and put them in ice or ice water so they can be prepared later. Many delicious food dishes can be made from the heart and the liver of the pig.

Splitting and head removal

Once the pig's organs are removed, you should wash the inside of the carcass before splitting it. With your handsaw, you should start splitting the carcass from the interior, between the hams. You should try to keep the split as close to the center of the backbone as possible. Saw through the tail area to a point halfway through the loin. Move around to the back of the pig, and keep sawing through the shoulder and neck to the base of the head. If the split goes off center, you should continue to saw through to the next vertebra and then get back to the center.

You will need to remove the pig's head at the **atlas joint**. This is the joint closest to the head. If you have properly split the carcass, then this joint should be exposed. After you have cut through this joint, you should continue to cut downward along the jawbone. Leave the jowls attached to the carcass. You can remove the tongue if you like, wash it completely, and place it with the liver and heart to be used later.

Next, you should remove the kidneys and the leaf fat. You can remove the leaf fat by loosening it from the diaphragm muscle and lifting up on it. You may have to scoop some of the leaf fat out with your hands. Leaf fat is used to make lard and is highly prized by many people.

Once you have finished splitting the carcass and removing the parts, you should examine the carcass and the organs to make sure everything looks satisfactory. This is normally a meat inspector's job, but as you are slaughtering your pig at home, you will have to perform this duty yourself. Look for bruises, injuries, parasites, abscesses, and tumors. Is there any congestion or inflammation in the lungs? Do the intestines, kidneys, and other organs look all right? Does the interior of the carcass look normal? You should check the carcass thoroughly for any signs of problems. If you do find something, it could affect the meat you intend to eat, or it could indicate that there is a possible sickness that could affect your other pigs. If you find something that concerns you, you should contact a meat inspector or a qualified veterinarian to take a look at the carcass.

The carcass will need to be thoroughly washed and chilled for 24 to 48 hours before meat can be cut.

Chilling the Carcass

Bacteria have already contaminated slaughtered hog carcasses during the slaughter process. These bacteria can spoil the meat unless their growth is stopped as soon as the slaughter process is finished. This bacterial growth can be drastically slowed by immediately chilling the carcass and keeping it at low temperatures. If you slaughter your pig during cold weather and the temperature is between 28 and 35 degrees, it is possible to wrap the carcass in a sheet, hang it, and chill it in a well-aired shed, as long as it is safe from any cats, dogs, or other predators looking for free meat. But, you must not allow the carcass to freeze. Freezing the carcass within a day following death can cause the meat to become tough.

If the weather is warm and you cannot cool the meat to less than 40 degrees, you will need to arrange to transport the carcass to a local market or butcher so it can be chilled for 24 to 48 hours. The carcass must be chilled for this length of time prior to making any cuts on the meat. Otherwise, bacterial growth cannot be prevented. No cuts of meat can be made on carcasses that have not been properly chilled because it is not safe to eat meat from carcasses that have not been properly chilled.

Marketing your Meat

As already discussed, you cannot sell any meat you slaughter on your own farm yourself. If you have meat that has been butchered by a licensed butcher for resale, then you can sell your meat to your chosen market.

Selling points

There are many different approaches you can take to marketing your pork products. You will need to decide what kind of market you are aiming for. Do you have heritage breeds? Then, you can capitalize on their slower growth and special flavors. Do you raise your pigs using organic practices? You can use that as a selling point. Are your pigs pasture raised? You can emphasize that point. Whatever it is that makes your farm and your pigs special, that is something you can use to sell your products. Make your products stand out so people will notice them.

Labeling

We have already discussed the kind of packaging that the butcher can provide. You will also need to consider the labels that will go on your packages. Are you artistic? Is someone in your family artistic? Can you come up with a nice logo for your farm? You will need an image or something else so people can identify a label with your products and your farm. It should be simple but easy to recognize. If you have a word or a catchy phrase, people may associate it with your products. Try to think of something simple that will stay in people's minds. Your phrase or logo does not have to come from an expensive ad agency. You and your family can think of a good idea for your products.

Who is your customer?

It always helps to ask yourself who your customer is. Who will buy your products? Are you selling to other people like yourself? Are you selling to chefs? To food lovers? Are you selling to suburban families? To local people? To people far away? The better you can picture and identify the person you are selling your pork products to, the easier it will be for you to understand how to market your products to them. If you know who you are selling to, you will know what they want. For example, you could sell your cuts of meats locally at farmers markets, or you could sell to someone who lives in the suburbs. They might like to buy a half hog, already dressed and ready to go in their freezer. The suburban family may not have the room or the time to raise their own pigs, but they enjoy pork products, and they have a large freezer to store the pork. Or, you could have your own website to sell your products to people interested in purchasing home-raised pork. It is possible for you to sell your pork in all of these ways. You simply have to know how to market your products. Local buyers might care more about buying pork that has been raised locally, and a suburban family might care more about your prices. Someone buying on the Internet might be more interested in the kind of pigs you have and how they are raised — pasture raised, organic practices, and so on. Learn to emphasize what your customers are looking for.

Where are you selling?

Are you selling at farmers markets? Local, independent grocery stores? A chain of grocery stores? Online markets? Gourmet or organic stores? Your own website? Are you selling directly to restaurants? To ethnic markets? How are you selling your products? How are other people selling their products? These are all outlets you may wish to consider when you look for outlets for selling your meats. As a small pork producer, you may not be able to compete with large producers in terms of lower costs or large volume, but you can find niche markets where your pork will be appreciated for its special qualities. Emphasize those points, and you can be highly successful.

Summary

In this chapter you learned how to slaughter your pig and considered how to prepare your pig for slaughter. You examined what to look for in a licensed butcher and processing facility and how to make arrangements to use their services. You also considered home butchering options: scalding and scraping, and skinning the pig. Finally, you looked at the things you need to consider in order to successfully market your pork products.

In the next chapter you will look at smokehouses. Many people are interested in smoking their own meat. You will consider different kinds of smokers, as well as whether you should buy a smoker, convert an old refrigerator, or build a permanent smokehouse. You will also examine health and safety concerns.

CHAPTER 11:

The Smokehouse

In the previous chapter, you learned how to slaughter your pigs, whether you take your pigs to a licensed processing facility or you prefer to slaughter them yourself at home. If you slaughter your own pigs or if you arrange with a licensed butcher to leave your meat fresh and uncured, you will need to know how to cure and smoke your own meat. In order to cure and smoke your own meat, you will need to have a smokehouse or a smoker of your own.

Curing Meats

Curing and smoking meats are age-old ways of preserving food. In previous centuries, before refrigeration, the processes of curing and smoking meats allowed people to safely keep meats for up to two or three years after slaughtering an animal. Meats have been cured and smoked since ancient times, and Native Americans would hang meat high in their teepees in order to smoke the meat from the fire below.

The primary agents involved in curing meats are traditionally salt and sugars, such as honey, corn syrup solids, and maple syrup. Sugar helps remove some of the salty flavor of the meat from the salt cure. Sugar also feeds beneficial bacteria in the meat. Salt dehydrates the meat, which helps to prevent the growth of bacteria. It encourages beneficial bacteria to grow, such as lactobacillus acidophilus. It also lowers the pH of the meat to 4.5, which is conducive to preventing harmful bacteria from growing. Salt also slows the oxidation process in the meat, which means it stops the meat from becoming rancid. Commercial meat cures include nitrites and nitrates used to prevent botulism poisoning, though these compounds are somewhat controversial, and the amounts used are carefully regulated in the United States. Some studies suggest that nitrites may cause an increased risk of cancer. There are other sources of nitrites and nitrates in the diet besides cured meats, such as fresh vegetables. It is not possible to cure meats without using nitrites to prevent botulism.

Smoking meats gives the meat flavor plus the smoke is also an antimicrobial and an antioxidant. The smoke helps prevent bacterial growth and oxidation to further prevent spoiling. It also gives the meat a characteristic "smoked" color and flavor.

Hams, bacon, and sausages are the meats usually cured and/or smoked. The exact recipes for curing the meats will vary from butcher to butcher, or farmer to farmer, according to taste. The bible for curing meats in the United States for the last century has been the Morton Salt Home Meat Curing Guide produced by the Morton Salt Company, which is available at **http://morton.elsstore.com/view/category/178-meat-curing/**. Morton makes a number of products for curing meats. If you are interested in curing your own meat, begin with this book and these products as your guide. You can make any changes once you are familiar with the process.

You should not use regular table salt, canning salt, or any other kind of salt for curing meat besides salt for curing meats. You will get salted meat, but you will not get properly cured meat. The color and flavor of the meat will not be properly developed. Likewise, you should not use salt for curing meat for cooking, baking, or table use. You must use curing salt for curing meats in order to have the correct recipe for preventing meat spoilage and to develop the right color and flavor of the meat. You can purchase curing salts at kitchen supply stores, gourmet shops, spice stores, and in some grocery stores.

Smoking Meats

Smoking meats has been practiced in the United States since colonial times. There are numerous old smokehouses still standing from the 18th century at Colonial Williamsburg, for example. Smokehouses tended to be small outbuildings near the kitchen, which during the time was also a separate building where hams could be hung to dry and smoke for months or even for a year or two. Some smokehouses were designed so a pit for fire was dug yards away from the smokehouse, covered up, and the smoke was channeled underground into the smokehouse. Some smokehouses were two-stories high to allow the smoke to circulate more fully. The home chimney was also used to hang meat and smoke it in some homes. Some smokehouses had a pit fire in the center of the one-story building. The fire was lit in the morning and allowed to go out during the day, with the smoke filling the room. It would be relit the next morning. These one-story smokehouses could be round or small square buildings, brick or log.

When it comes to smoking meats, there are many different approaches. You will need to consider some basic things about your own situation before deciding to smoke your own meat. How much meat do you have to smoke? Are you smoking the meat from one pig or half a dozen? If you are only smoking the meat from one pig, then it probably will not make sense to go to the trouble and expense to build a smokehouse. But, if you will smoke the meat from half a dozen pigs each year, then you will need a lot of room to smoke your meat. In that case, you may need your own smokehouse.

Types of Smokers

It is a good idea to know what kind of smokers are available before you make any decisions. It is unlikely you will need professional quality smokers. But, if you would like top-of-the-line smokehouse equipment, you can visit a site like SausageMaker.com (**www.sausagemaker.com**) where they sell smokehouses and meat smokers.

- A modest propane gas smoker that is 36 inches high with multiple shelves sells for $300.
- A 20-pound stainless steel insulated smokehouse sells for $500.
- A digital 30-pound country-style insulated smokehouse sells for $650.
- A 50-pound stainless steel smokehouse sells for $1,500.
- A 100-pound gas smokehouse that is stainless steel inside and out sells for $4,655. A similar smokehouse that is electric sells for $4,420.

SausageMaker.com also sells accessories for smokers, including different kinds of wood chips, such as applewood, hickory, and mesquite.

Homestead Harvest is another site to check for smokehouses. They also have component kits to build your own smokers. An electric component kit is $700. A gas component kit is about $1,000 for a 100-pound smoker. A gas smoker heating assembly element is $350. This is the price of the heating assembly if you have all of the other parts. The "assembly" is what the heating element is called.

There is a professionally made smoker for just about any kind of meat smoking need you may have. You can find used smokers like these for sale on Craigslist (**www.craigslist.org**) and some other listing sites.

Buying a Production Smoker

A production smoker, or a commercial meat smoker, has a capacity from 100 to 1,500 pounds. If you intend to buy a commercial meat smoker, then you will probably be doing so for business purposes. If you intend to sell the meat you process, you will need to become licensed and inspected to butcher your own pigs.

Commercial meat smokers can be either natural gas, electric, or pellet burning. Electric models are most commonly used for indoor kitchens. You will need to make sure you have direct venting and a good range hood to provide exhaust for the smoke.

A model with a digital electronic control will be able to better control the smoking and monitor the temperature of the meat. Some models come with convection fans to keep the smoke moving inside the smoker. Rotating racks can also keep the meat moving and reduce "hot spots," or the places in an oven than can get overly hot. Look for a well-insulated commercial smoker because this will reduce your production costs.

Choose a commercial smoker that is tightly sealed and well ventilated. Less ventilation results in less shrinkage of the meat and more product to sell.

Barbecue Smokers

Barbecue smokers are not used to smoke meats following butchering but instead are used prior to cooking meats. This is called "hot smoking" as opposed to the cold smoking used for smoking hams and bacon. Barbecue smokers come in three styles: horizontal smokers, offset smokers, and vertical smokers.

- The horizontal smoker is normally cylindrical in shape. It is barrel shaped or drum shaped.

- If the horizontal smoker has a fire pit attached near the end of the pit, then it is called an offset smoker. The offset smoker keeps the meat away from direct heat, giving the meat a smoky flavor without grilling it. This style is often used by professional barbecue people or those who enter barbecue contests.

- Vertical smokers are also cylindrical, but they stand up instead of lie down. They can also be rectangular. They are often made of heavier material than the other smokers. Instead of having a firebox, the fire may burn on a tray or in the bottom of the smoker. There is often more direct heat reaching the meat.

You can make good barbecue with any of these styles, and they all come with a wide array of options and features depending on your preferences and budget.

Converting an Old Refrigerator

If you are simply interested in smoking the meat from one pig or you want to smoke cuts of meat you have purchased, many people are happy having an old refrigerator as a smoker. A refrigerator or old chest freezer is relatively small compared to a smokehouse, and it may not seem to be able to hold much meat at one time. However, a refrigerator or chest freezer can be operated 24 hours per day, which increases their usefulness.

1. In order to convert an old refrigerator or chest freezer into a small smokehouse, you will need to remove all of the rubber gaskets, as well as the plastic molding. The motor and the compressor must also be removed.

2. It is best to use an older model with a steel interior, but most modern models have plastic interiors, and you must take these plastic interiors out prior to use. A 10- to 12-cubic-foot refrigerator model can make a very good small smokehouse.

3. Once you remove the plastic molding, you will need to install metal sheeting on the interior walls and on the inside of the door. You can purchase metal sheeting from a hardware store, such as Ace Hardware® (**www.acehardware.com**) for about $40 for a 24-foot-by-24-foot section of aluminum sheet metal. Or, you can reuse some old sheet metal if you have any lying around.

4. Cut holes near the bottom and side at the top of the refrigerator for your pipes. You will need both an intake and an exhaust pipe. You can use masonry pipe for the exhaust. The intake pipe should be a masonry pipe that is 4 to 6 inches in diameter.

5. Seal the pipes with furnace cement. Expect smoke and heat to come out around the door edges with the rubber gasket removed.

6. Anchor the refrigerator firmly in place. You can use cinder blocks to bolster the refrigeration or freezer by placing them against the unit. Leave space at the top of the enclosure for your hams to hang. You can use bricks to space your racks inside. Bricks will retain heat at night if you decide to let the fire die down.

7. Drill a hole in the door, place a meat thermometer in it, and cement it in place so you can tell at all times what the temperature is inside. It should be around 130 degrees for hams and bacon.

8. Your firebox is simply the area where the wood is burned. It should be made of firebricks or concrete, and it should have a small, 2-inch intake vent leading into the refrigerator. The firebox does not need to be large. It may only be 1½ square feet because you will only need to have a small fire. Ensure the masonry pipe is securely sealed in place. The masonry pipe will vent your smoke into your refrigerator or freezer. Ideally, your firebox should be dug several feet underground and at least a yard away from the refrigerator.

9. You will need to have a connecting pipe that angles toward the refrigerator from the underground firebox. It is best to place the firebox behind or to the side of the refrigerator so it will not interfere with your access to the door.

10. Once completed, pack the area with sand, dirt, and cement, leaving only a small access door to the firebox so you can add wood and tend the fire.

Building a Permanent Smokehouse

If you plan to smoke meats year after year, then you may prefer to build your own smokehouse. Building your own smokehouse is not as difficult as you may believe. You will need to choose a location that is not too close to your house because most people do not want to have smoke blowing into their homes or to have their belongings take on the odor of smoke.

If you intend to smoke hams, bacon, sausages and if you will have more than a few of these meats, then it is best to build a larger smokehouse. A smokehouse does not need to be complex or expensive to build. If you can hammer, you will probably be able to build one.

1. Start by building a simple box-type structure comprised of four sides. Your smokehouse does not necessarily need to have a floor, but you will need a roof and a door. You will need to provide good insulation for the building, though. One-inch thick fiberglass insulation works well.

2. Once the building is constructed, you should cut a hole about 12 inches from the bottom of the ground for a 3-inch vent in the side of the building. Make a similar hole on the opposite wall at the top of the building where a metal pipe can be fitted, and then put a cap on the pipe.

3. At this point, place racks, shelves, and hanging hooks in your smokehouse for your meats.

4. The last thing you need to do is to make a firebox for your smokehouse, similar to the underground firebox described for the converted refrigerator. Connect the firebox to the opening that you

made into the smokehouse that is 12 inches above ground with a pipe. You control the temperature in your smokehouse by opening and closing the vent at the top of the building. The temperature should be able to reach up to 170 to 200 degrees, though you will usually keep the temperature lower for slow drying your hams and other meats.

5. Alternately, instead of a firebox in the ground, you could connect your lower pipe to a drum you keep stoked for smoking.

You can also make an old-fashioned style smokehouse by building a small wooden structure around a cinder block fire pit, with vents at the top of the building.

When building your smokehouse, you will need to check with your local government agencies about zoning regulations. There may be zoning issues about how far the building must be from your property line or house and what materials you can use.

Health and Safety

As you may imagine, when working with meat and fires, it is necessary to exercise a great deal of caution. Even though the purpose of the smokehouse or smoker is to smoke meat, smokers and smokehouses need to have ventilation and a means of controlling their temperature. Be sure that pipes, fireboxes, and vents are properly constructed and working as they should.

If you are curing meat, follow directions. Meat needs to be correctly cured to prevent harmful bacteria from growing. Otherwise, you run the risk of eating bad meat. Make sure you allow plenty of time for the meat to cure and smoke. On the other hand, do not leave hams exposed to smoke for too long. Creosote from the smoke may begin to build up on the meat. **Creosote** is a black, gummy, tar-like substance that can build up inside a smokehouse from the smoke. It can also build up on hams and other cured meats if they are exposed to smoke for a long time. The creosote will give the meats a burned flavor. Creosote from wood smoke is an insecticide, a fungicide, and a preservative. It is not as harmful as creosote from coal, but it is still not something you particularly want to have on your meat. If a smokehouse is too tightly constructed and humidity is not allowed to escape, it can build up on the meat, producing mold. None of these things will ruin your meat, but you will have to cut off the creosote or mold, meaning you will have to throw away some of your meat.

Summary

In this chapter, you have looked at some of the reasons why meat is cured and smoked. You have also considered some of the different kinds of smokers available. The chapter also included instructions for converting an old refrigerator into a smoker and for building a small smokehouse. Additionally, you looked at some important health and safety issues for cured and smoked meats.

Conclusion

The Complete Guide to Raising Pigs: Everything You Need to Know Explained Simply has covered the basics of raising pigs, from farrowing piglets to smoking your own meat. After reading this book, you should be ready to get a pair of young pigs and start having fun getting to know these wonderful animals. If you already have a few pigs, this book should provide a continual reference as you raise them through different stages. Even experienced pig raisers should find new and helpful information in this book.

Although most pork in the United States is produced by large producers, there is an increasing number of small pig raisers and people who are interested in homesteading. These pig raisers may simply want to raise pigs so they can fill their own freezers with pork, but many of them are finding excellent niche markets for flavorful pork from heritage breeds, pork that is slow grown and pasture raised, or pork that is raised using more natural methods. If you are one of these small pig raisers, then this book is designed for you. It should provide you with plenty of good information to help you succeed.

Good luck, and enjoy your pigs.

Appendix

Breed Associations

American Berkshire Association
2637 Yeager Road
West Lafayette, IN 47906
765-497-3618
www.americanberkshire.com

American Landrace Association
Member of the National Swine
Registry
P.O. Box 2417
West Lafayette, IN 47996
765-463-3594
www.nationalswine.com

American Mulefoot Hog
Association and Registry
18995 V Drive
Tekonsha, MI 49092
517-767-4729
http://mulefootpigs.tripod.com/

American Yorkshire Club
Member of the National Swine
Registry
P.O. Box 2417
West Lafayette, IN 47996
765-463-3594
www.nationalswine.com

Chester White Registry
Member of Certified Pedigreed
Swine (CPS)
P.O. Box 9758
Peoria, IL 61612
309-691-0151
www.cpsswine.com

Hampshire Swine Registry
Member of the National Swine
Registry
P.O. Box 2417
West Lafayette, IN 47996
765-463-3594
www.nationalswine.com

North American Large Black Pig
Registry
c/o Ted Smith
Stillmeadow Farm
740 Lower Myrick Road
Laurel, MS 39440
601-426-2264
stillmeadow@c-gate.net
www.albc-usa.org/cpl/
largeblack.html

National Hereford Hog Record
Association
c/o Ruby Schrecengost, Secretary
22405 480th Street
Flandreau, SD 57028
605-997-2116

National Spotted Swine Record,
Inc.
Member of Certified Pedigreed
Swine (CPS)
PO Box 9758
Peoria, IL 61612
309-691-0151
www.cpsswine.com

Poland China Record Association
Member of Certified Pedigreed
Swine (CPS)
P.O. Box 9758
Peoria, IL 61612
309-691-0151
www.cpsswine.com

Red Wattle Hog Association
c/o Josh Wendland, President
21901 Mayday Road
Barnes, KS 66933
785-944-3574
wendland@ourtownusa.net

Tamworth Swine Association
c/o Shirley Brattain
621 N CR 850 W
Greencastle, IN 46135
765-794-0203

United Duroc Swine Registry
Member of the National Swine
Registry
P.O. Box 2417
West Lafayette, IN 47996
765-463-3594
www.nationalswine.com

Additional Organizations

American Livestock Breeds
Conservancy
P.O. Box 477
Pittsboro, NC 27312
919-542-5704
www.albc-usa.org

National Association of Animal
Breeders (NAAB)
P.O. Box 1033
Columbia, MO 65205
573-445-4406
www.naab-css.org

FFA
P.O. Box 68960, 6060 FFA Drive
Indianapolis, IN 46282
317-802-6060
www.ffa.org

National Pork Producers Council
122 C Street NW
Suite 875
Washington, D.C. 20001
202-347-3600
www.nppc.org

4-H Club
1400 Independence Avenue SW,
Stop 2225
Washington, D.C. 20250
202-720-2908
www.4-h.org

Glossary

Agalactia: a lack of milk in sows.

All-in and all-out method: a method of pig production where all sows due to give birth within a few days of each other are brought to the farrowing area at the same time. Once the piglets are born and are ready to wean, they can all go out at the same time.

Amino acids: compounds necessary to for building muscle, for gestation, for lactation, and for growth.

Ampulla: the halfway point where the sperm and egg meet for fertilization.

Anthelmintics: another term for medication that kills all worms present in an animal or prevents a worm infestation.

Atlas joint: the joint closest to the head.

Barrow: a young male pig.

Boar taint: the foul taste of testosterone included in boar meat.

Boar: an adult male pig.

Butcher-hog: a hog weighing between 220 and 260 pounds, raised for slaughter.

Creep: the area in a farrowing site that is off limits to the sow where piglets can get feed.

Creosote: a black, gummy, tar-like substance that can build up inside a smokehouse and can also build up on hams and other cured meats if they are exposed to smoke for long periods.

Crossbred: a pig that is a cross of different breeds of pig.

Cutability: a larger percentage of the pig's mass translates into marketable cuts of meat.

Dam: mother.

Driving: the term used to describe moving pigs.

Drylot: a large, dry lot on dirt where pigs are kept.

Earmarks: a traditional method of marking animals where the farmer makes notches according to a system and chart with the litter marking on one ear and the individual pig marking on the other ear.

Estrus: the time when the female is receptive to the boar.

Evisceration: removing a pig's organs and intestines.

Farrow to finish: raising amounts from birth to being market or slaughter ready.

Farrow: when a pig gives birth.

Feed efficiency: the amount of feed it takes for a pig to gain 1 pound.

Feeder pig: a young pig, normally just weaned, that a breeder produces, but someone else raises; between 35 and 70 pounds.

Finished hog: a fattened hog ready for market or slaughter.

Finishing: when you try to put on the final pounds before sending a hog to market.

Follicles: the hair-like structures that cover the surface of the ovaries.

Gilt: a young female pig.

Grower pig: a young pig being raised for market or for slaughter; a pig more than 50 pounds.

Grower to finish: raising pigs from the weaner stage to market or slaughter size.

Heart girth: the distance around the pig just behind its elbows.

Heritage breeds: swine breeds that have a long history, sometimes going back several hundred years.

Hog panels: woven-wire panels that are easy to move and attach to metal posts.

Hog: a pig weighing more than 120 pounds.

Homeopathy: a natural approach to medicine that utilizes plants and minerals.

Leaf fat: the fat around the kidney that is used to make lard.

Loafing shed: a simple lean-to shed pigs can use to get out of the rain and weather.

Market hog: a pig that weighs 220 to 260 pounds and could be sold at market.

Mastitis: hardened and painful teats that can be so painful a sow refuses to feed.

Metritis: an infection in a sow's uterus.

Oviduct: the tube that connects the uterus and ovaries; this is where the egg is fertilized.

Piebald: a color of pig that is black with white bands or spots and points.

Pig board: a tall board with a handle that is used to direct a pig's movements.

Prepuce: a pouch of skin that hangs under a boar's stomach.

Proven sow: a sow that has already produced champion offspring.

Purebred: a pig that belongs to a recognized pure breed of pig.

Scurf: the layer of skin, oil, and cells on the skin.

Seedstock herds: purebred herds with breeding stock available for others to purchase.

Shoat: a pig from weaning age to 120 pounds.

Sire service: a stud service available for a fee.

Sire: father.

Slow food movement: an international movement that has grown as an alternative to the fast food that people eat on the go.

Snare: a flexible lead that loops around the pig's upper jaw and snout.

Sow: an adult female pig.

Swine: generic term for pigs.

Table pigs: pigs you can raise for your own table or freezer.

Tagging: putting a small tag with an ID number through a pig's ear.

Tankage: a liquid product from rendered animal carcasses that offers pigs a good source of phosphorus and calcium.

Thumps: diaphragm spasms that are indicative of serious anemia in pigs.

Topline: the spine or back of the big, especially in silhouette.

Type: the essence of a breed that distinguishes one breed of pig from another.

Underline: the line formed along the stomach of the pig.

Vas deferens: a muscular tube located behind the testes that helps propel sperm into the urethra.

Wattles: long pieces of flesh that hang from a pig's cheeks.

Bibliography

Hasheider, Philip. *How To Raise Pigs*. Minneapolis, Minnesota: Voyageur Press, 2008.

Klober, Kelly. *Storey's Guide To Raising Pigs*. North Adams, Massachusetts: Storey Publishing, 2009.

McFarlen, Arie B. *Pigs: Keeping a Small-Scale Herd for Pleasure and Profit*. Laguna Hills, California: Hobby Farm Press, 2008.

Van Loon, Dirk. *Small-Scale Pig Raising*. Pownal, Vermont: Storey Publishing, 1978.

Author Biography

C A R L O T T A   C O O P E R

Carlotta Cooper was born and raised in Tennessee. Her grandparents were farmers, and she grew up with horses, dogs, and other animals, including some pigs raised for the table. Her grandparents  had their own smokehouse and cured their own meats. She attended the University of the South in Sewanee where she graduated with a B.A. in English as class salutatorian. She attended graduate school at the University of Virginia, studying English literature, and did graduate work in writing and rhetoric at the University of Tennessee at Chattanooga.

Professionally, Carlotta is a freelance writer, specializing in writing about animals. She has been breeding and showing dogs for more than 20 years. She is a contributing editor for the dog show magazine *Dog News*. She lives in the middle of farm country in Tennessee now, writing about veterinary issues, animal reproduction, genetics, and raising and caring for animals. Carlotta is now thinking of adding some Gloucestershire Old Spots to her home menagerie.

Index